AREA 9

THE OUTER DOMAIN

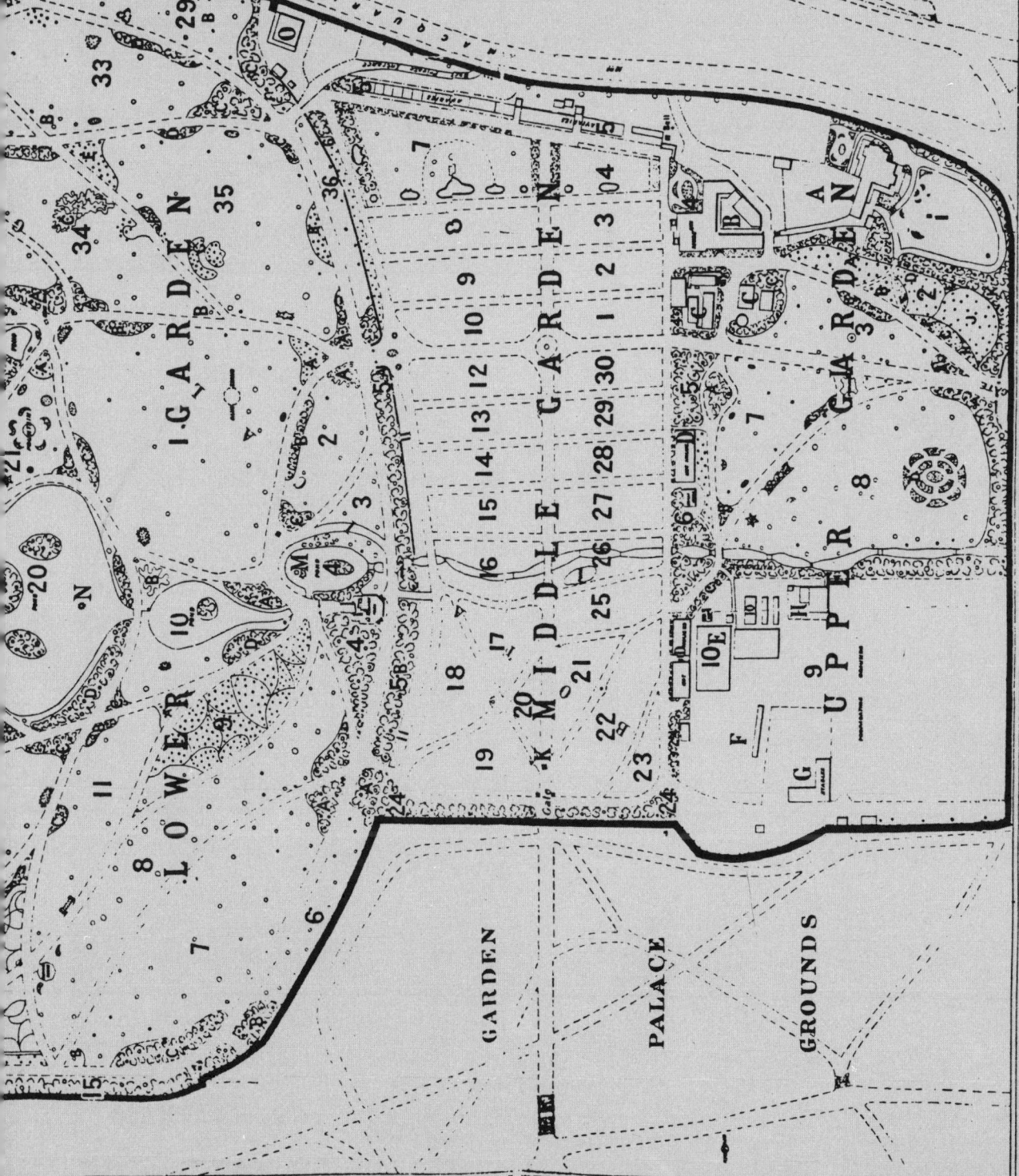

I. G A R D E N

L O W E R

M I D D L E   G A R D E N

U P P E R   G A R D E N

G A R D E N

P A L A C E

G R O U N D S

Abou

THE ROYAL
BOTANIC GARDENS,
SYDNEY

Joseph Henry Maiden, ISO, FRS, FLS
1859–1925
Government Botanist and Director of the Sydney Botanic Gardens 1896–1924
an untiring and dedicated botanist who fully appreciated the value
of the history of his science.
(Royal Botanic Gardens Library)

THE ROYAL
BOTANIC GARDENS,
SYDNEY

A History 1816-1985

Lionel Gilbert

Melbourne
Oxford University Press
Oxford Auckland New York

OXFORD UNIVERSITY PRESS
Oxford New York Toronto
Delhi Bombay Calcutta Madras Karachi
Singapore Hong Kong Tokyo
Nairobi Dar es Salaam Cape Town
Melbourne Auckland
and associates in
Beirut Berlin Ibadan Nicosia

National Library of Australia
Cataloguing-in-Publication data:

Gilbert, Lionel A. (Lionel Arthur), 1924– .
   The Royal Botanic Gardens, Sydney.

   Includes index.
   ISBN 0 19 554719 5.

   1. Royal Botanic Gardens (Sydney, N.S.W.) – History.
   2. Botanical gardens – New South Wales – Sydney –
   History. I. Title.

580′.74′49441

Edited by Lee White
Designed by Guy Mirabella
Typeset by Asco Trade Typesetting Ltd., Hong Kong
Printed by Kyodo Shing Loong Printing, Singapore
Published by Oxford University Press, 7 Bowen Crescent, Melbourne
OXFORD is a trademark of Oxford University Press

Contents

Foreword by
the Hon. Neville Wran, QC, MP,
Premier of NSW *vii*

Preface by
Dr L.A.S. Johnson, Director,
Royal Botanic Gardens, Sydney *ix*

Acknowledgements *xiii*

Introduction *1*

1 Background and Beginning *4*

2 Nine Acres in Corn 1788–1810 *9*

3 The Governor's Demesne 1810–21 *17*

4 A Scientific Institution? 1821–48 *40*

5 Crisis and Consolidation 1848–96 *75*

6 The Maiden Era 1896–1924 *114*

7 Old Objectives, New Directions 1924–85 *142*

8 Beyond the Garden Wall *168*

9 Retrospect and Prospect *177*

Sources *189*

Abbreviations *190*

References *191*

Index *207*

Continents have therefore, at all times, been remarkable for their immense botanical treasures; and, among the rest, that of New-Holland, so lately examined by Messrs. BANKS and SOLANDER, rewarded their labors so plentifully that one of its harbors obtained a name suitable to this circumstance (Botany Bay).

JOHN REINOLD FORSTER:
*Observations made during
a Voyage round the World . . . ,*
London, 1778

In all the British colonies, I believe Botanic Gardens have been considered as necessary appendages; although the purposes to which these should be applied have either been uninvestigated, or altogether lost sight of. At present, they form elegant and delightful promenades for the ladies and the burgesses; they provide delicious fruits for the desert [sic] of the Governor, or for those who are honored by an entree to the Lady Governess's bourdoir [sic]. As secondary objects, they may have in view the transmission of the fairer flowers of the mountain, to deck the green-houses of the noble or illustrious, or the collection of the meaner productions of the wilderness to lumber the works of science, or to record the existence of those whose names would otherwise have been soon consigned to oblivion.

JOHN HENDERSON: *Observations
on the Colonies of New South Wales
and Van Diemen's Land,*
Calcutta, 1832

Our Botanic Garden 'growed' like Topsy, following at first the economic requirements of a colony formed under peculiar circumstances, and as regards the garden, we became scientific in spite of ourselves . . .

JOSEPH HENRY MAIDEN to the
Royal Society of New South
Wales, 1 May 1912

A true botanic garden is a collection of scientifically-labelled plant specimens from known locations, planted in some regular order, and supported by a herbarium, a collection of similar plant material that has been dried and preserved for scientific purposes. In fact the garden is a museum of living specimens while the herbarium is a museum of preserved specimens.

RONALD G. HODGES in *Australian
Parks and Recreation,* August 1979

Foreword

The Royal Botanic Gardens of Sydney are a living link with the history of Sydney and indeed Australia. Dr Gilbert's history, beginning in 1788, is therefore very welcome as a permanent record of their eventful past.

The traditional foundation date of the Gardens is regarded as being the 13th June 1816 and for that reason the Gardens' 170th birthday has been chosen as an appropriate launching date for this book. This is not to say that the events before 1816 were not as significant in the history of this area of land. It was the site of the first farm for the colony and as such can claim to be the birthplace of Australia's primary industry.

This book traces the events since then to the present time when the Gardens, with the National Herbarium of New South Wales, have become acknowledged as a centre of scientific endeavour in a place of great beauty, high in the affections of the citizens of Sydney and admired by visitors.

In an atmosphere of concern for the environment, it is appropriate that the Royal Botanic Gardens should continue to pursue and encourage wider understanding of the role of plants. The Royal Botanic Gardens and Domain Trust, established by the present New South Wales Government, is aware of this need and its efforts in this direction will ensure its place in the future history.

THE HON. NEVILLE WRAN, QC, MP,
Premier of NSW

Preface

Australia's Aborigines were hunters and gatherers. When their displacement by Europeans began almost two hundred years ago, the newcomers immediately instituted their own cultural basis: cultivation of useful plants, and soon of ornamentals. Woccanmagully became Farm Cove.

The hungry, phosphorus-deficient soils of Port Jackson's shores were no place for a farm, as Governor Phillip soon learned. But, after some uncertain years, cultivation continued and, under Governor Lachlan Macquarie, the place became a Botanic Garden—the third in the Southern Hemisphere.

The Garden gradually acquired plurality of title, as additional areas were brought in from the encompassing Domain. At first largely an entrepot through which 'useful' plants came and went, to and from Australia, it soon acquired some scientific and amenity functions.

High drama has not marked the history of the Gardens, unlike that of our sister institution, the Australian Museum, which saw a (probably justifiably) recalcitrant Director carried out in his chair to the footpath, or of the Melbourne Gardens, where the notable scientist Ferdinand von Mueller was sacked because of his unyielding principles. Botanists and horticulturists, though, are not always mild, and some fierce internal and external battles and feuds have occurred. Antagonism from commercial nurserymen nearly wrecked the institution in its worst period in the 1830s and '40s, and crassness and interference drove Allan Cunningham out in disgust. His remains rest in the Gardens, and the Cunningham Building honours his memory.

The everlasting Charles Moore, in forty-eight years as Director, survived politically contrived malice, showed a Committee of Management to be superfluous in his time, and developed the landscape and living collections to a level of some excellence. Then he sat back and largely rested on his laurels for the later years of his tenure.

Real botany had to wait for Joseph Henry Maiden, whose vigour, insight and ability raised the Gardens to a peak. Notably through his own researches, Maiden put the Gardens (and the somewhat pretentiously designated 'National' Herbarium of New South Wales) on the world map of systematic botany. Melbourne had enjoyed such a position under Mueller; when he was ousted the Botanic Garden by the Yarra became beautiful but intellectually void. Systematic botany in Victoria has been a long time recovering.

Maiden was of a broader mind than Mueller, and a defter politician. He perceived that his Gardens had to serve the State (and indeed the world) in divers ways, and he met that challenge in all of them.

A dismal slough followed these heights, from 1925 to the 1940s, and a lesson is to be learnt from it.

Doubtless, Ministers and departmental heads considered it a sound management decision to review and 'rationalize' the Botanic Gardens to fit in with the perceived objectives and functions of the Department of Agriculture. The ensuing stultification leads one to reflect on how domination by a well-meaning but bureaucratic department destroyed the soul of an organization whose aims and scope it did not understand, at first under a genial but ineffectual Director who did not understand them either, and later under no Director at all.

Working with his official masters, but never allowing himself to be browbeaten by them, Maiden had run his own ship, fought his own battles, held to and developed his own concepts and ideas. *Après lui, le déluge:* no science of quality, lapse of educational and community functions, over-emphasis on bread-and-butter rural botany, loss of all thematic meaning in the Gardens. Hawkey, the Curator in the split administration of the 1930s, defended the Gardens against a justified charge of deterioration; his grounds were that grass-cutting and pruning were in good order! Such a purblind attitude lasted in horticultural areas in the Gardens until very recent times, and has brought new internal battles, but today's policies, in a more modern version, are a return to Maiden's philosophy.

Robert Henry Anderson began the painful and slow revival, but mostly on the scientific side. A humdrum scientist himself, he recognized the need for quality. Helped and encouraged by Joyce Winifred Vickery, he built up a competent botanical staff, and the Herbarium's recovery was under way. The only real improvement in the Gardens was also botanical, a program of accurate labelling under George Chippendale, aided by his botanist colleagues. A landscape team was developed and a few attractive plantings were made, but philosophy was lacking and co-ordination of themes with interpretation, education and research was virtually unknown until the 1970s.

Anderson fought valiantly to defend the Gardens against incursions, but could not stop the ugly intrusive expressway. The political and bureaucratic climate of the 1940s, 1950s, and 1960s cannot be praised. Anderson was burdened by this, by his own lack of experience outside Australia and even New South Wales, and by the worldwide intellectual mediocrity that characterized systematic botany and botanic gardens development in the first half of the twentieth century. It is thus remarkable that he left his institution so much better than he found it, and he is justly commemorated by the Anderson Building, the old Herbarium he strove unavailingly to have replaced.

His successor, Knowles Mair, continued to fight for a new Herbarium building, with support from Roy Watts, Director-General of Agriculture, but otherwise still in a deadening official and political climate. He achieved three major advances: full administrative reunification of the scientific and horticultural-cum-amenity arms of the now Royal Botanic Gardens, appointment of the first ecologist, and construction of the pyramid tropical glasshouse. The gift of land for a satellite garden at Mount Tomah was also welcomed by Mair.

We have arrived at recent times, and I shall not eschew a personal note. Encouraged at first by Knowles Mair to apply for the Directorship on his retirement, I decided not to do so. I had seen Anderson's and Mair's strivings so often defeated by the powers, and had observed that they spent most of their time divorced from science in tedious dealings with the outdoor areas, which scarcely improved, as well as with the intrinsically important but incongruous millstone, Centennial Park.

Several months as Acting Director began to change my mind and recalled Anderson's wisdom: 'I do it my way; you will do it in yours'. Meanwhile, Dr John Beard had been appointed. For the first time since 1925 a scientist of some world standing was Director.

Beard was positive and vigorous, and his stirring of the pot has paid off in more recent events. Unfortunately, his personality and approach were not calculated to fit in with the departmental environment, or indeed with the attitudes of some of his staff. In retrospect, one's sympathies can be divided. After eighteen months, Dr Beard's connections with the Gardens were severed.

His initiation of an education service came to fruition, his Botanic Gardens Society foundered but the Friends of the Gardens became a reality a decade later. Separation from the Department of Agriculture would have gratified him. His visions of a grandiose fountain in Farm Cove, with waterside restaurants and nocturnal entertainment, and of an incongruous and trendy Japanese Garden, will, I hope, remain in the limbo to which they have been relegated.

My appointment followed, and subsequent developments can be traced in this book and elsewhere. As I near a retirement forced by present Government policies on age, I pay tribute to those who left the Gardens better than they found them, namely Fraser, Moore, Maiden, Anderson and Mair, and to those who did not do that, for reasons of incompatibility with prevailing forces, but who shook things up, with benefit felt in later times. Such were Cunningham and Beard. Many others within the institution have had profound effects for good upon it; their names are in this book and in present staff listings. Governors, and more recently Governments, have ranged from beastly to beneficent, and in the present day the Gardens have flourished under a generally sympathetic Government and a Trust with whom it is a pleasure to work.

The Royal Botanic Gardens and Domain Trust, through its successive Chairmen, The Hon. Sir Alexander Beattie, Professor Michael Pitman, OBE, and Mr John Ferris, AM, have welcomed the production of a comprehensive history of the Gardens.

In this book, Dr Lionel Gilbert gives us a fluent and scholarly historian's detailed narrative, which I have read with fascination. I have taken this opportunity to summarize and set in context the progress, and at times the regress, of the Gardens as seen by a Director. Sometimes this is a different viewpoint.

L.A.S. JOHNSON
Director
21 May 1985

Acknowledgements

Probably I should never have contemplated this project had I not experienced the good fortune to be introduced to the study of botany by Miss Thistle Y. Harris (later Mrs David Stead) at Sydney Teachers' College in 1942. Nor have I forgotten her kindness in writing a biology course for a class for one to enable an essential 'option study' to be completed by external study during the unsettled years of 1943–45.

Studies in field botany inevitably led to visits to the National Herbarium in the Sydney Gardens in 1945–46, when I was encouraged by Dr Douglas Cross, Dr Mary Tindale and the late Rev. H.M.R. Rupp. The two last mentioned became frequent correspondents and patient instructors, so that visits to the Herbarium and to the Gardens in general became more pleasant and enlightening.

Twenty years ago and more, there was further encouragement from Professor Noel Beadle and Professor Russel Ward of the University of New England during a long period of studying and writing botanical history. During the last five years, the Gardens Director, Dr Lawrie Johnson and his deputy, Dr Barbara Briggs, have been most co-operative and helpful, not only making available many resources and facilities but also reading the entire manuscript and offering valuable suggestions. I am also grateful for the interest and practical assistance of the Gardens Librarian, Ms Anna Hallett, of Messrs David Bedford and Andrew Mitchell, and of other current staff members including my former colleague, the Extension Officer, Mr Edwin Wilson, who will appreciate that the guide and information book produced under his editorship has been of considerable help in writing a work which does not purport to be a guidebook.

Former Directors Mr Knowles Mair and Dr John Beard favoured me with reminiscences and frank and informative answers to many queries, as did Mr George Chippendale. Mrs Helen Nicholas and Mr Harry Craig, grandchildren of J.H. Maiden, and Mrs Beverley Passey, daughter of R.H. Anderson, kindly made available family papers which proved invaluable. Mrs Lorna Backhouse received me generously and hospitably when I sought information about the late Alfred and Effie Brunet, and I was similarly received by Mrs Phyllis Anderson, widow of the former Director, by Assistant Director, Mr Warwick Watson, during his period of convalescence, and by Mrs Dorothea Cox, great niece of Charles Moore.

There was ample reason to be grateful to Mr T.J. Best, whose survey of records relating to the Gardens and Domains, compiled as a student assignment, greatly facilitated the search for material in the State Archives Office, where the Principal Archivist, Mr D.J. Cross, the Assistant Principal Archivist, Mrs D. Troy and Senior Archivist, Mrs C.M. Shergold and other staff members were invariably courteous and helpful. Ms Baiba Berzins, Ms Jennifer Broomhead, Ms Elizabeth Imashev and their colleagues made frequent work sessions in the Mitchell Library both enjoyable and informative, while Mrs Pam Ray, Ms Barbara Perry and Ms Sylvia Carr ensured that less frequent visits to the National Library in Canberra were equally profitable. Dr Peter Stanbury, Mr Alan Davies, Ms Catherine Snowden and Ms Alison Lea gave freely of their time and knowledge during and after my introduction to the magnificent resources of the Historic Photographs Collection in the Macleay Museum at the University of Sydney. The interest and help of Ms Helen Cohn, Librarian, Royal Botanic Gardens, Melbourne, are likewise thankfully acknowledged.

I am grateful also to the Premier, the Hon. Neville Wran, for his interest and his Foreword, and to the Secretary of the Premier's Department, Mr Gerald Gleeson, last chairman of the former Gardens Trustees, for helpful information and advice. Similarly the encouragement of the present Trust, under successive chairmen, Professor Michael Pitman, OBE, and Mr John Ferris, AM, is warmly appreciated.

Finally, thanks are due to Miss Shirley Dawson and her assistants for their careful preparation of many of the illustrations, and to Mrs Mary McClenaghan, who transformed an unorthodox and much amended typescript into something clearly presentable.

LIONEL GILBERT
Armidale
1 May 1985

Introduction

There are three things which have stimulated men through the ages to travel far and wide over the surface of the globe, and these are gold, spices and drugs. It is to the two latter of these universal needs of man that we may trace the origin and foundation of botanic gardens.

ARTHUR W. HILL
Royal Botanic Gardens, Kew
1914[1]

The Sydney Botanic Gardens are an important part of a world heritage. As urban populations and concern for the environment increase together, hardly a day passes without news of water or air pollution, provision of ample open space, encroachment on shrinking green belts, threats to natural habitats, and the likely extinction of certain plant and animal species. We also hear much about 'the knowledge explosion', new research techniques, retrieval systems, the profitable use of leisure, active and passive recreation and the 'quality of life'. In this context, the Botanic Gardens of Sydney, like many elsewhere, have assumed an even greater significance than they hitherto enjoyed. When the Bicentenary of European settlement of Australia is celebrated in 1988, that significance will be still more fully appreciated.

The reasons for establishing, developing and maintaining botanic gardens throughout the world, sometimes over several centuries, have long been considered. In the general sense, gardens can be traced, with some ease and clear certainty, to the ancient civilizations of the Nile, the Tigris and the Euphrates, of ancient Greece and Rome, of ancient China and Japan, and of medieval and renaissance Europe. There is pictorial evidence of the garden of Nebamun, a scribe of Thebes, about 1400 BC;[2] there is literary evidence of King Nebuchadnezzar's 'hanging gardens' of Babylon and of King Darius's gardens in ancient Persia; there is archaeological evidence of well-ordered, if small, gardens in Pompeii flourishing when Vesuvius smothered them in 79 AD. There are Biblical references to gardens. Genesis records that the Garden of Eden (Garden of Pleasure) was mankind's first place of residence and sustenance, while elsewhere the Old Testament records royal gardens, cultivations enclosed by walls and hedges, and a vineyard which was converted into 'a garden of herbs'. The New Testament records Christ's betrayal in the Garden of Gethsemane (Garden of the Oil Press) and burial in a garden wherein Joseph of Arimathaea had had a tomb hewn in the rock.

Horticultural historians have described sacred and mystic groves, 'paradise gardens', fantastic, formal, classical and romantic gardens, Arcadian gardens—and botanic gardens. The more venturesome have proceeded beyond recorded history to Neolithic times to proclaim that 'the inventors of gardening' were 'almost certainly' New Stone Age women, whose crops 'were certainly all edible, and, a little later, medicinal'. Much later, urban civilizations were to develop 'ornamental gardening'.[3]

It has been suggested that the garden of the Greek philosopher Theophrastus (c. 370–286 BC), pupil of Plato and Aristotle, 'may claim ... to be the first botanic garden' for it was 'a place of study for his friends and disciples'.[4] On the other hand, it has been claimed that the first botanic garden where there was organized study of plants was established by the Aztec ruler, Moctezuma I (c. 1390–1469).[5]

The traditional European view is that the first botanic gardens, in the modern sense, were founded during the mid-sixteenth century in Italy, at Padua and Pisa, Florence and Bologna. Others followed at Leipzig, Leiden and Heidelberg, and in 1621 at Oxford.[6] In mediaeval and renaissance Europe gardens were often attached to monasteries and universities for the cultivation of 'simples' or medicinal herbs and other useful plants. Extensive gardens were also established around royal palaces and noble homes, often assuming a scientific form and function.

Although far removed in time and space from the gardens of antiquity, it will be seen that Sydney's Gardens share much of this common tradition. Today, the combined role of botanic gardens is generally agreed to be scientific, educational, aesthetic and recreational,[7] with the scientific aspect the most important, if perhaps the least appreciated.

The Royal Botanic Gardens of Sydney occupy some 30 hectares (75 acres) on a matchless harbourside location on Port Jackson. The site has long been admired. In 1969, a visitor with a special eye for botanic gardens, declared:

> ... I know of no botanical garden in the world so beautifully sited as Sydney's, out on a peninsula in a grand bay of which Trollope wrote a century ago, 'I despair of being able to convey to my reader any idea of the beauty of Sydney harbour. I have seen nothing equal to it in the way of land-locked sea-scenery ...'

Insulated by about 34 hectares (85 acres) of adjoining Domain parkland, the Gardens form a remarkable and delightful environmental contrast to the enveloping, and often threatening, metropolis, with its three million inhabitants, its one and a half million motor vehicles, and its ever-soaring architecture.

For nearly two hundred years, the area occupied by Sydney's Botanic Gardens has enjoyed, or merely endured, the attention of countless people from all levels of society and all walks of life—dutiful governors and humble farmers, disgruntled convicts and furtive trespassers, shrewd administrators and canny politicians, resourceful directors and industrious superintendents, watchful gardeners and eager educationists, painstaking nurserymen and enthusiastic horticulturists, benign tree-lovers and hawk-eyed bird-watchers, dedicated scientists and hedonistic picnickers, casual sight-seers and pensive pedestrians, cooing couples and somnolent sunbathers, spirited children and desperate parents—and, most recently, earnest environmentalists, stalwart joggers and zealous agents of the National Estate. Long, long before them all, the local Guringai people enjoyed and utilized the place for food-gathering, ceremonial occasions and other aspects of tribal life.

In their various ways, people have found sustenance, knowledge and enlightenment in this place; a quiet and beautiful retreat; a leafy refuge from the summer heat; a paragon of landscape art; an inspiration to reflect or to write, to paint or to photograph; a source of fresh and fragrant air untainted by carbon monoxide. The rather colourless modern expression 'a place of passive recreation' falls far short of conveying the real nature, purpose and significance of Sydney's Botanic Gardens. They are, in the restrained phrases of our observant visitor:

> ... a monument to the origins of Australian horticulture and agriculture; botanically a good teaching and research institute; and horticulturally an example of very good gardening.[8]

Established during the reign of King George III, the Gardens were granted the royal epithet in 1959 by his great-great-great-great-granddaughter, Queen Elizabeth II, who thereby bestowed an appropriate sign of patronage on Australia's oldest scientific institution.

In order to appreciate something of the fascinating story of the foundation and development of these Gardens, and indeed, the appropriateness of their royal patronage, we need to look back beyond the beginnings of European settlement in Australia to the celebrated age of world-wide maritime exploration and scientific discovery two centuries and more before man set foot on the moon.

Chapter 1

BACKGROUND
AND BEGINNING

I believe no age can boast a more rapid progress in the Knowledge and History of Nature than the present; the States of Europe vye with each other in pursuit of Information, and the great Academies have sent their Disciples to all Quarters of the Globe to explore the animal and vegetable World, to gratify the Thirst for Knowledge, & enrich their Parent Countries with objects of ornament and Utility ...

It is now determined to send a great Body of Persons who have forfeited their Liberty here to the Southern Hemisphere and to plant them on the Banks of Botany Bay on a Soil and in a Latitude fitted for the Culture of every Thing that can satisfy the wants and Indulge the Appetites of Men ...

'R.H.' to SIR JOSEPH BANKS
13 December 1786[1]

During the first century or so of its 'official' existence, the Royal Society of London for Improving Natural Knowledge, granted a charter by King Charles II in 1662, clearly demonstrated the link between knowledge and influence. Among the earliest concerns of the Society were the circulation of the blood; the Copernican theory which held that revolving planets followed orbits around the sun; the nature of comets, stars and sun-spots; the improvement of the telescope through the better grinding of lenses; 'the descent of heavy bodies, and the degrees of acceleration therein; and in divers other things of like nature'.[2]

Favoured by royal patronage and comprising some of the foremost scientists in the kingdom, it was natural that the Society should have been consulted by the government for specialist advice on a varied range of subjects including changes in the calendar, the ventilation of prisons, the protection of buildings and ships from lightning, the determination of latitude, and the comparison of British and French standards of linear measure. In 1710 the Society assumed control of the Royal Observatory at Greenwich.

In 1761 the Royal Society persuaded the British Government to send astronomers to two vantage points, St Helena and Sumatra, in order to observe the predicted transit of the planet Venus across the sun. Cloudy weather spoiled the observations at St Helena, while the vessel bound for Sumatra was attacked by the French, and no observations were made. This double disaster made it all the more imperative that observations of the next predicted transit in 1769 should be successful. After all, whatever obstacles may have been presented by capricious clouds and fractious Frenchmen, it had been an Englishman, the Reverend Jeremiah Horrocks (1619–41), a Lancashire curate, who predicted such astronomical phenomena, and national pride was at stake.

Young King George III, the Society's patron from his accession in 1760, was so impressed by the Society's memorial of 12 November 1767 that he granted £4000 towards the project. Accordingly, in March 1768 a Whitby-built collier, the *Earl of Pembroke*, was commissioned as His Majesty's bark *Endeavour*, which was extensively refitted for a long scientific voyage and placed in command of Lieutenant James Cook, RN.

The proposed expedition immediately caught the imagination of the scientific fraternity in Britain, more especially of the Fellows of the Royal Society, among whom Mr Joseph Banks, at the age of twenty-three, had just taken his place as one of the youngest ever elected. He was later to serve the Society as president for forty-two years.

The Society urged that 'Joseph Banks ... a Gentleman of large fortune ... well versed in natural history' should be permitted to join the *Endeavour* 'with his Suite'. It was a considerable 'suite', including two naturalists, Daniel Solander and Herman Spöring; two landscape and natural history artists, Alexander Buchan and Sydney Parkinson; two tenants from Banks's estate and two negro servants. Only four were to return—Banks himself, Solander and the two tenants.[3]

Those engaged in developing the Royal Gardens at Kew were also greatly interested in this expedition and in its possible botanical results—Augusta, Dowager Princess of Wales; John Stuart, third Earl of Bute; William Aiton, 'His Majesty's Principal Gardener at Kew' since 1759; and of course George III himself. On the Continent, probably no one had a greater interest in the expedition than the celebrated Carolus Linnaeus or Carl von Linné, Professor of Botany at the University of Uppsala since 1742, former mentor of Solander and correspondent of Spöring.

When the expedition was about to sail via Cape Horn for the observation point of Tahiti, John Ellis, a Fellow of the Royal Society, informed Linnaeus that 'no people ever went to sea better fitted out for the purpose of Natural History or more elegantly', and that the 'expedition would cost Mr Banks ten thousand pounds'.[4]

The *Endeavour* sailed from Plymouth in August 1768 with nearly 100 men after the final ingredient of a classic adventure story had been added in the form of a packet of secret instructions to be opened once the astronomical observations at Tahiti had been completed. The transit was duly observed, and the *Endeavour*, agreeable to the additional instructions, sailed south and west, to New Zealand and on to 'fall in with the Coast of New Holland' which was faithfully charted from Point Hicks to Cape York.

On 3 May 1770, during a week of feverish exploring and collecting around the landing-place which Cook was finally to name Botany Bay, Banks noted: 'Our collection of Plants was now grown so immensly [sic] large that it was necessary that some extraordinary care should be taken of them least they should spoil...'[5] Banks accordingly spent the day taking ashore about 200 quires of drying paper and spreading the plant specimens on a spare sail to dry in the sun. Now, 215 years later, many of these specimens have an honoured place in the Herbarium of the Royal Botanic Gardens, Sydney.

After its remarkable voyage around the world which lasted nearly three years, the *Endeavour* returned to England in July 1771 with a sadly depleted complement. Banks and Solander were immediately fêted. In August they were presented to the King, and in November Oxford University bestowed upon them honorary doctorates of Civil Law. The huge collections of plants, seeds, shells, insects, reptiles, birds, bottled and dried specimens, native implements, paintings, sketches and reams of notes were transferred to Banks's London house in Burlington Street, where Solander was installed as librarian, cataloguer and secretary. There was talk of a great illustrated botanical work to be published. Linnaeus was elated and suggested that the newly explored land in the south be called 'Banksia'.[6] Banks himself, however, disappointed the scientific world by deciding almost immediately to embark on other, perhaps greater, ventures of discovery, a decision which Linnaeus said 'almost entirely' deprived him of sleep.

In 1772 the Dowager Princess of Wales died, and her son, King George III, thereupon dispensed with the services of the horticulturally efficient but politically unpopular Earl of Bute, and appointed Banks as his special adviser on matters relating to the Royal Gardens at Kew. Thus began a new phase in Banks's influence in natural science and in government, and a friendship which later sustained the King— a friendship which ended only when both men died in 1820. It has been estimated that during George III's reign from 1760 to 1820, some 7000 hitherto unknown plants were introduced into England,[7] some directly by Banks himself, most by Banksian collectors acting for Banks in the interests of Kew.

Linnaeus died in 1778, still disappointed that the scientific results of the *Endeavour* had not been properly written up and published; Cook was killed in Hawaii in 1779; Solander suffered a stroke and died in 1782. Banks, President of the Royal Society since a stormy election in 1778 and a baronet since 1781, came to be regarded as something of an oracle in matters of natural philosophy. He was consulted upon plans for exploration, for settlement, and for developing botanic gardens; he was asked to advise on plant diseases, earthquakes, wool production, leather tanning and currency.

In 1779 Banks was called to give evidence before a House of Commons Select Committee investigating the possibility of forming overseas colonies. Of the Australian east coast he said:

... the Proportion of rich Soil was small in Comparison to the barren, but sufficient to support a very large Number of People ... there were no Beasts of Prey ... our Oxen or Sheep, if carried there, would thrive and increase; there was great Plenty of Fish ... The Grass was long and luxuriant, and there were some eatable Vegetables, particularly a Sort of Wild Spinage; the Country was well supplied with Water; there was abundance of Timber and Fuel, sufficient for any Number of Buildings.[8]

Nevertheless, he advised that intending settlers would need to take a year's supply of food, beverages and clothing as well as implements, seeds and livestock.

To his journal of August 1770, Banks had confided:

For the whole lengh of coast which we saild along there was a sameness to be observd in the face of the countrey very uncommon; Barren it may justly be calld and in a very high degree, that at least that we saw. The Soil in general is sandy and very light: on it grows grass tall enough but thin sett, and trees of a tolerable size, never however near together ... upon the Whole the fertile soil Bears no kind of Proportion to that which seems by nature doomed to everlasting Barrenness.

Water is here a scarce article or at least was so while we were there ...

A Soil so barren and at the same time intirely void of the helps derivd from cultivation could not be supposed to yeild much towards the support of man.[9]

Like a promising wine, Banks's views had mellowed considerably during the intervening nine years. He strongly advocated Botany Bay as a place for settlement, thereby supporting the plan of a former shipmate, James Maria Matra, who at one stage understood Botany Bay was to be settled under Banks's own direction. Encouraged by Banks's public stand, the British government decided to establish a penal colony in Australia which was intended to be largely self-sufficient.

If King George III's 'trusty and well-beloved Arthur Phillip, Esq. ... Captain General and Governor-in-Chief ... of New South Wales' had had his way, an advance party would have arrived at the proposed site of the Botany Bay settlement 'two or three months' before the bulk of the fleet comprising storeships and transports. There would have been the opportunity to select a suitable location with anchorage facilities and a water supply, huts could have been built, stores secured, cultivations established, defences constructed and preparations made to receive the large number of settlers who would almost inevitably arrive stricken with scurvy or dysentery after so long at sea.

However, the captain was obliged to supervise preparations for the large and unprecedented expedition. He was also obliged to wage a paper war with those responsible for drafting the convicts and victualling the ships; he had to ensure that the vessels were cleaned and fumigated, that the convicts were adequately clothed and that there were ample supplies of medicines and ammunition. He noted that the provision of six scythes indicated dim prospects of cultivation in the new colony, and that the provision of only five dozen razors promised little for the grooming of the men.

Having obliged Lord Sydney and Mr Under Secretary Nepean with some final reminders of sundry bungles and dealt with a seamen's pay dispute, Phillip decided to sail. At day-break on Sunday, 13 May 1787, 'The Sirius made the Signal for the whole fleet to get under Way'[10] and the eleven vessels moved from Spithead unhonoured, unsung and wellnigh unnoticed, although a clergyman on the Isle of Wight recorded that on seeing the fleet, he 'prayed for its safety, and for the souls of the prisoners'.[11]

Maintaining his intention to reach Botany Bay ahead of the fleet, Phillip, on 25 November 1787 when 'only 80 leagues' from Cape Town, 'left the Sirius and went on board the Supply tender'.[12] The *Supply*, however, 'sailed but very indifferently' and when Phillip arrived in Botany Bay on 18 January 1788 the remainder of the fleet was hard on his heels, with three ships arriving on the 19th and the remainder on the 20th.

The Governor did not share Banks and Solander's botanical enthusiasm of eighteen years before; he considered Botany Bay was exposed, damp, unhealthy and generally unsuitable for European settlement. Having given instructions for the preparation of a landing place, Phillip proceeded to examine the inlet marked on Cook's chart as Port Jackson. Here he 'had the satisfaction to find one of the finest harbours in the world, in which a thousand sail of the line might ride in perfect security'.[13] The 'different coves of this harbour were examined' and the spot chosen was 'one which had the finest spring of water, and in which ships can anchor close to the shore ... In honour of Lord Sydney, the Governor distinguished it by the name of Sydney Cove'.[14]

Having made his decision, Phillip ordered the fleet to move to the new site. After three of the vessels had collided in the manoeuvre, the fleet came round to Port Jackson. The collisions were as embarrassing as alarming, for two French ships under Jean François Galaup, Comte de la Pérouse, had anchored in Botany Bay before the near-tragic drama was enacted. There was some disquiet over this incident, 'Every one blaming the Rashness of the Governor in insisting upon the fleets workg. out in such weather, & all agreed it was next to a Miracle that some of the Ships were not lost, the danger was so very great'.[15] But then came the reward—the sight of Port Jackson, declared to be 'one of the finest harbours in the world—I never Saw any like it—the River Thames is not to be mentioned to it and that I thought was the finest in the world—this Said Port Jackson is the most beautiful place ...'[16] The fleet reassembled in Sydney Cove, where

In the evening of the 26th the colours were displayed on shore, and the Governor, with several of his principal officers and others, assembled round the flag-staff, drank the king's health, and success to the settlement, with all that display of form which on such occasions is esteemed propitious, because it enlivens the spirits, and fills the imagination with pleasant presages. From this time to the end of the first week in February all was hurry and exertion.[17]

Part of that exertion was to investigate another cove, a little to the east, and to clear the ground where another 'spring of water' flowed into Port Jackson. As far as the settlers could tell, the Guringai people called the cove 'Woccanmagully', but the new arrivals called it Farm Cove. Much later, when the land behind that cove had long since been converted into Sydney's Botanic Gardens, the area was described as 'classic ground'.

Chapter 2

NINE ACRES IN CORN
1788–1810

The wild appearance of land entirely untouched by cultivation, the close and perplexed growing of trees, interrupted now and then by barren spots, bare rocks, or spaces overgrown with weeds, flowers, flowering shrubs, or underwood, scattered and mingled in the most promiscuous manner, are the first objects that present themselves.

GOVERNOR ARTHUR PHILLIP, 1788[1]

Agreement about the scenic grandeur of Port Jackson was matched by concern about the capacity of the country to sustain the proposed settlement. Surgeon Arthur Bowes Smyth's impressions of Botany Bay proved to be equally applicable to Sydney Cove:

Upon the first sight one wd. be induced to think this a most fertile spot, as there are great Nos. of very large & lofty trees, reachg. almost to the water's edge, & every vacant spot between the trees appears to be cover'd wt. verdure: but upon a nearer inspection, the grass is found long & coarse, the trees very large & in general hollow & the wood itself fit for no purposes of buildg. or anything but the fire—[2]

Surgeon George Worgan agreed:

On approaching the Land ... It suggests to the Imagination Ideas of luxuriant Vegetation and rural Scenery ... gentle risings & Depressions, beautifully clothed with variety of Verdures of Evergreens, forming dense Thickets, & lofty Trees appearing above these again, and now & then a pleasant checquered Glade opens to your View—Here, a romantic rocky, craggy Precipice ... There, a soft vivid-green, shady Lawn ...

Happy were it for the Colony, if these Appearances did not prove so delusive as upon a nearer Examination they are found to do ...

Such country would 'necessarily require much Time and Labour to cultivate any considerable Space of Land together ...'[3]

Phillip despaired of being able to give Lord Sydney 'a just idea' of the amount of labour required to clear the ground for cultivation:

The necks of land that form the different coves, and near the water for some distance, are in general so rocky that it is surprizing such large trees should find sufficient nourishment, but the soil between the rocks is good, and the summits of the rocks, as well as the whole country round us, with few exceptions, are covered with trees, most of which are so large that the removing them off the ground after they are cut down is the greatest part of the labour.[4]

Some observers, like Midshipman Daniel Southwell, sanguinely believed that 'we may form some judgement of the quality of the soil by the immense quantity of wood it produces',[5] but others were inconsolable. Major Robert Ross, the Lieutenant-Governor, advised Mr Under Secretary Nepean of his total disillusionment:

I do not scruple to pronounce that in the whole world there is not a worse country than what we have yet seen of this. All that is contiguous to us is so very barren and forbidding that it may with truth be said here nature is reversed: and if not so, she is nearly worn out ...[6]

Even Lieutenant Ralph Clark, who had so extolled the beauty of Port Jackson in January 1788, was convinced by July that this was 'the poorest country in the World without Exception' with 'not a Quarter of an Acre of clear ground ... to be Seen ...' and within ten days he used the same description in at least three letters to his friends in England.[7] Even five years later, Thomas Watling, the convict artist, wrote in similar vein, taking the opportunity to question the assessments of the late James Cook:

The flattering appearance of nature may be offered as the best apology for those mistaken eulogiums lavished by a late eminent circumnavigator upon this place... The face of the country is deceitful; having every appearance of fertility; and yet productive of no one article in itself fit for the support of mankind...[8]

Despite the stark reality of the situation and the prevalence of gloom, Governor Phillip set about his daunting task with energy and optimism. His detailed instructions bade him attend to many diverse matters including the exploration of the coast and the transmission, 'with all convenient speed', of 'a report of the actual state and quality of the soil at and near the said intended settlement, the probable and most effectual means of improving and cultivating the same'.[9] He was also to cultivate the flax plant, *Phormium tenax*, which Banks had noted in New Zealand, and believed would 'be a great acquisition to England', and to colonize Norfolk Island.

By mid-May 1788, when Phillip composed his first and remarkably long despatch to Lord Sydney, there was much to report. Exploration had been carried out and Norfolk Island, where Banks's flax had been found, had been settled under Lieutenant Philip Gidley King whose party had included 'two men who understood the cultivation of flax'. Phillip drew attention to his botanical difficulties:

... the flax-plant described by Capt. Cook I have never met with, nor had the botanists that accompanied Mons. La Perouse found it when I saw them, and which was some time after they arrived: and here, my Lord, I must beg leave to observe, with regret, that being myself without the smallest knowledge of botany, I am without one botanist, or even an intelligent gardener, in the colony; it is not therefore in my power to give more than a very superficial account of the produce of this country, which has such variety of plants that I cannot, with all my ignorance, help being convinced that it merits the attention of the naturalist and the botanist.[10]

This was not quite fair to Phillip's personal servant, Henry Edward Dodd, who must have been 'an intelligent gardener' since he was entrusted with the establishment of a small farm at 'a spot at the head of the adjoining cove' where some land was cleared in February[11] on the site of the Botanic Gardens.

In view of Sir Joseph Banks's special interest in the colony, it seems strange that he was unable to have a qualified botanist, horticulturist or agriculturist included in the First Fleet's official establishment. Ironically, the French expedition in Botany Bay was much better prepared for making botanical investigations. Phillip's emissary Lieutenant P.G. King found that La Pérouse had 'besides yᵉ Astronomer Monsieur Dagelet ... a very capital Botanist from yᵉ Jardin du roi called de la Martinniere also a draftsman' and 'an Abbe who is ... a collector of Natural Curiosities'.[12] The complement of *La Boussole* and *L'Astrolabe* included a Franciscan friar, Abbé le Receveur, chaplain and naturalist; the Prévosts, uncle and nephew, botanical draftsmen; Bossieu de la Martinière, medical officer and botanist; Dufresne, naturalist, and Collignon, gardener and botanist. In talent and number they comprised a strong scientific contingent. Le Receveur died during the six-week stay at Botany Bay, and his grave may still be seen at La Perouse. The expedition shortly came to grief on Vanikoro Reef, Santa Cruz, and all were lost.

Meanwhile, amid the 'hurry and exertion' of establishing the new settlement, Australian agriculture, horticulture and fruit growing were born more or less simultaneously. Seeds and living plants from England, Rio de Janeiro and the Cape were landed, and some were transferred to 'some ground ... prepared near his excellency's house on the East side' of the Tank Stream, where the colonists 'soon had the satisfaction of seeing the grape, the fig, the orange, the pear, and the apple, the delicious fruits of the Old, taking root and establishing themselves in our New World'.[13] Other species procured on the voyage out included coffee, cocoa, cotton, bamboo, jalap, banana, lemon, guava, quince, strawberry, prickly pear (or 'cochineal fig' employed for the production of a red dye used for uniforms), oaks and myrtles, and the 'Spanish Reed', that is, the European or Giant Reed, *Arundo donax*.[14] Some of these proposed introductions did not survive the voyage; others did not survive the dry, sandy, phosphorus-deficient soil of the new cultivations. As Surgeon Worgan noted in June 1788: 'The Plants which we brought here, from the Rio de Janeiro and the Cape of Good Hope look tolerably promising, for yᵉ most part, but some of these have perished, and others appear to be withering'.[15]

Australian agriculture began less than a kilometre to the east of the plantations near 'his excellency's house' at the scene of the labours of Henry Edward Dodd and his band of convicts 'on the spot of ground which was cleared soon after our arrival at the adjoining cove, since distinguished by the name of Farm Cove...'.[16] By July 1788 there were '9 Acres in Corn' (that is, cereals, chiefly wheat) at this 'spot of ground'. The convicts who cleared and sowed the pitifully limited area of alluvial soil on the eastern side of what was later called Botanic Gardens Creek unwittingly prepared for the development of one of the first botanic gardens outside of Europe. Thus began the earliest stage of an almost indiscernible transition from the agricultural Government Farm, to the horticultural and ultimately scientific Botanic Garden.

As early as April 1788 a convict, Francis Fowkes, sketched in a rather fanciful map the general plan of the settlement to show the Governor's temporary quarters flanked

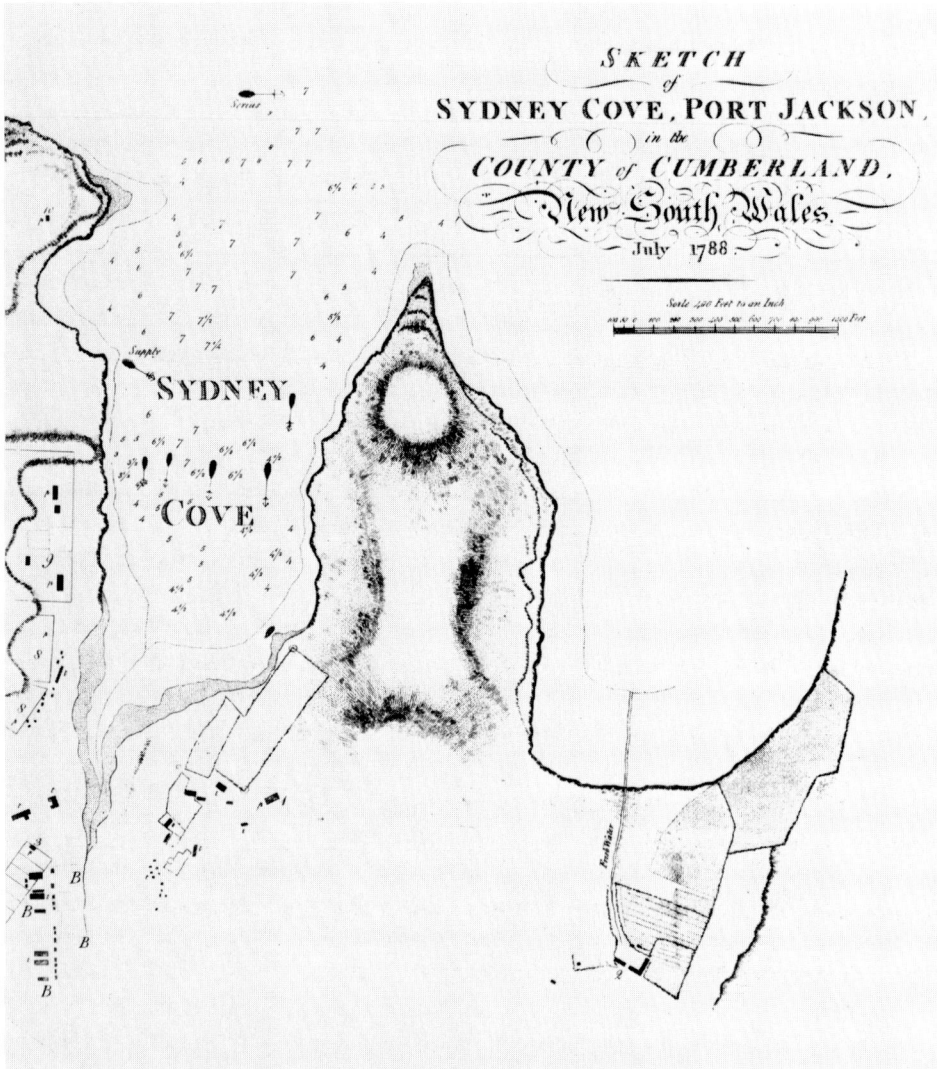

Lieutenant William Dawes's sketch map of 1788 showing the '9 Acres in Corn' on the eastern side of the fresh water stream, later to be called the Botanic Gardens Creek.
(The Voyage of Governor Phillip to Botany Bay, *London, 1789*)

by gardens to north and south, with the much larger 'Farm' on 'the adjoining cove'. Lieutenant William Dawes's professionally drawn map of July clearly showed 'A Farm: 9 Acres in Corn' across the creek and adjacent to the shady palm walks in the area long known as the Middle Garden within Sydney's Botanic Gardens. This Government Farm grew rather rapidly, and by September 1788 comprised 'six acres of wheat, eight of barley, and six of other grain, as raised on the public account, and in a very promising way'.[17] This semi-official account read well enough, but it was either premature, unduly sanguine, or gently exaggerated for public approval.

Writing to his brother in June 1788, Surgeon George Worgan made no attempt to paint a picture of any agricultural paradise in the South Seas to influence public opinion:

The Spots of Ground that we have cultivated for Gardens, have brought forth most of the Seeds that we put in soon after our Arrival here, and besides the common culinary plants, Indigo, Coffee, Ginger, Castor Nut, Oranges, Lemons & Limes, Firs & Oaks, have vegitated from Seed, but whether from any unfriendly, deleterious Quality of the Soil or the Season, nothing seems to flourish vigorously long, but they shoot suddenly after being put in the Ground, look green & luxuriant for a little Time, blossom early, fructify slowly & weakly, and ripen before they come to their proper Size. Indeed, many of the Plants wither long ere they arrive at these Periods of Growth... I opened one of my Potatoe Beds, & found 6 or 7 at each Root; Indian Corn, and English Wheat, I think promise very fair; But on the whole, it is evident, that from some Cause or other, tho' most of ye Seeds vegetate, the Plants degenerate in their Growth exceedingly.[18]

The same state of affairs apparently developed in the '9 Acres in Corn', for by early 1789 the comparative sterility of the Government Farm had been amply demonstrated. The Farm was 'still attended to, and the fences kept in repair', but 'there was not any intention of clearing more ground in that spot'.[19] Indeed by January 1789 'whatever expectations could be formed of successful cultivation' centered on Rose Hill, or Parramatta,[20] to which outpost Henry Edward Dodd was transferred in March to replace James Smith as superintendent of agriculture. Dedicated to the cause of agriculture to the end, Dodd's death on 28 January 1791 was 'accelerated by exposing himself in his shirt for three or four hours during the night, in search after some thieves who were plundering his garden'.[21] His tersely worded monument, the oldest in St John's cemetery at Parramatta, makes no mention of his pioneering work in agriculture at Farm Cove, nor of the fact that he had 'joined to much agricultural knowledge a perfect idea of the labour to be required from, and that might be performed by the convicts; and his figure was calculated to make the idle and the worthless shrink if he came near them...'.[22]

The fading of official hopes for a productive Government Farm at the head of Farm Cove was further indicated by the fact that between 1794 and 1807 private individuals were permitted to lease land around the Cove, including the original Government Farm itself. Those permitted to take up leases in and around the early Government Farm included Nicholas Divine, superintendent of convicts, who in 1794 leased 'Eight Acres known by the Name of The Governor's Old Farm'. This, with other land, he offered 'to be let or sold' early in 1804.[23] By 1800 Divine had blocks on either side of Botanic Gardens Creek, while the peninsula between Farm Cove and Woolloomooloo Bay had been leased to Governor John Hunter's servant, Nathaniel Franklin (who suicided in 1798); Alexander McDonald; James Callum and Frederick Markett.[24] By 1797 an additional block on the eastern bank of Botanic Gardens Creek had been leased to Thomas Alford, who in 1814 was described as 'an Old and very faithful Servant to Government, and who has been for upwards of Twenty five years in this Colony, the greater part of which time he had served as Head Government Gardener to the entire satisfaction of every Successive Governor'. Alford, a convict who became free by servitude, made application, with Governor Macquarie's support, for his wife Mary to join him in the colony after their quarter of a century of separation. However, in 1817 the Governor arranged for Alford to return to England as a reward for his long and faithful service to colonial horticulture.[25]

Another landholder in the Government Farm area was Joseph Gerrald, one of the 'Scottish martyrs' transported for sedition late in 1795. Governor Hunter permitted

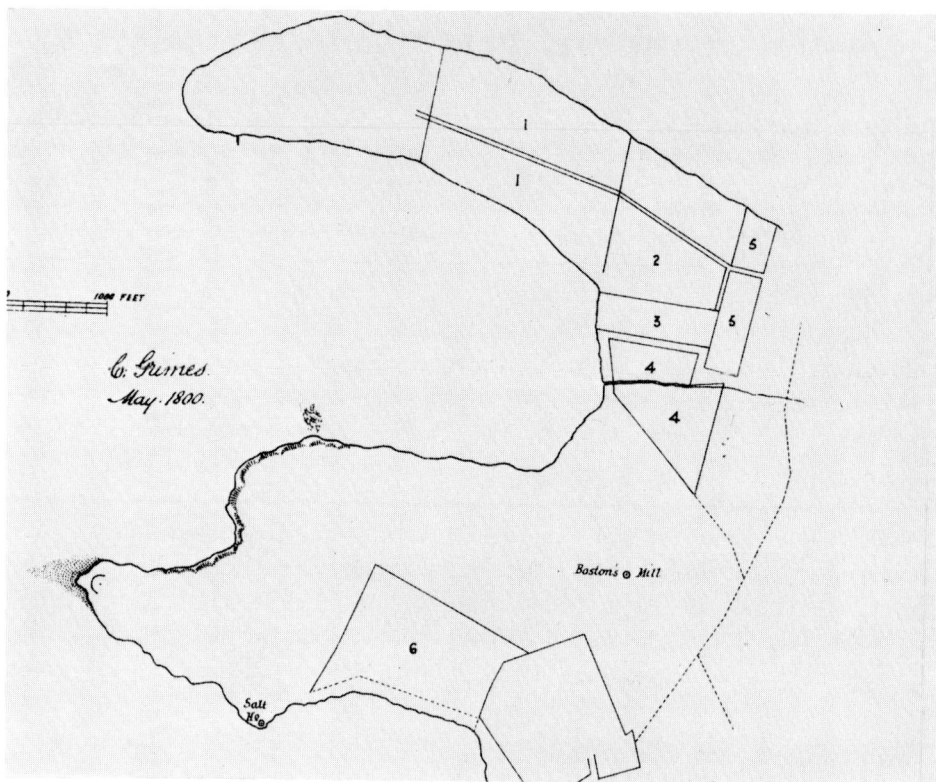

Part of Surveyor Charles Grimes's Plan of Sydney, May 1800, showing the Farm Cove blocks which were leased within the old Government Farm area.
(Historical Records of NSW, *Vol. V*)

Gerrald to live quietly at Farm Cove, but he soon died of advanced tuberculosis and at his express wish 'was buried in the garden of a little spot of ground which he had purchased at Farm Cove'. The traditional site of his grave, with its monument declaring that 'he died a martyr to the liberties of his country', was between a large Norfolk Island pine (the Old Wishing Tree) and Botanic Gardens Creek.[26]

These and other occupations were allowed in clear defiance of one of Governor Phillip's last official acts. On 2 December 1792 Phillip caused a line to be drawn from the head of Cockle Bay (later Darling Harbour) to the head of Garden Cove (later Woolloomooloo Bay) 'within which all the ground' was to be 'reserved for the Crown and for the use of the Town of Sydney'. It was 'the Orders of Government that no Ground within the Boundary line is ever granted or let on Lease…'. This provision effectively made the entire peninsula bounded by the western shore of Woolloomooloo Bay and the eastern shore of Darling Harbour a great Crown reserve, a point which became significant when law suits were pursued about 120 years later. Within this straight line (which incidentally transected the 100 acres granted to Mr Commissary John Palmer two months later) Phillip also marked out a shorter, more devious line for a ditch demarcating a much smaller area between the vicinity of Government House (near the present intersection of Bridge and Phillip Streets) more or less south-easterly to the head of Woolloomooloo Bay. This enclosed the area of the old Government Farm and the present Botanic Gardens.[27]

By 1807, in Governor William Bligh's time, land on the western and southern shores of Farm Cove was declared to be 'absolutely necessary for the use of Government House'. Despite 'leases improperly granted on it', this land was said to be 'now improving'. This 'improvement' may well have been descriptive of the state of tenure as well as of the condition of the land, as the Crown took steps to regain the area Phillip had delineated by a line for a ditch which was duly 'made by Governor Bligh'. This policy of resumption earned Bligh considerable notoriety.

In January 1807 Bligh warned 'all Persons whatever ... to keep their Hogs, Goats, or Cattle ... from off Government Farm, between the Ditch crossing from the back of Government House towards Woolloomoolla...'.[28] Six months later another proclamation reminded colonists of Phillip's 'lines of demarkation' within which all land was 'the property of the Crown'. Accordingly, 'a number of houses adjacent to Government House, to its great annoyance' then 'occupied by David Dickinson Mann, Abraham Ramsden, John Apsey, William Kimber, John Shea, Ferdinand Meurant, and others' were to be vacated and removed, perhaps to unoccupied land which might be granted in lieu of the prohibited areas.[29] As Deputy-Commissary Robert Fitz pointed out:

It appears that Gov'r Philip marked out these lines, within which no buildings were to be erected. However, Gov'r Hunter and Governor King thought differently, and permitted several persons to erect houses thereon, and the latter granted leases for 14 years to some... The only purpose to which the land thus obtained is the enlargement of the domain for the grazing of the Governor's horses, which before consisted of many acres—at least 150 ... if one Governor can do away the act of a former one, all property of whatever nature must be uncertain...[30]

One can certainly appreciate the problems of tenure during the foundation years; we can also appreciate how close we may have been to having no Crown reserves at all, and thereby, no Botanic Gardens on the present site.

Bligh clearly threw himself with characteristic energy into the implementation of his plans for improving the grounds around old Government House. Surgeon John Harris advised the wife of former Governor King:

He has so changed everything about Government House that you would be entirely lost in it... The shrubbery has also undergone a thorough change—no grass now growing in it, all laid out in walks with clumps of trees. Even the poor tomb of young Kent is anihilated [*sic*]. All the rocks in the garden is blown up and carried away. Not less than 80 or 90 men have been constantly employed since you went away for these purposes. Carriage roads are now all round Bennelong's Point, and down about Farm Cove, all ditch'd in and no thoroughfare allowed. All dogs are ordered to be shot, at which his secretary and body-guard have had much amusement...[31]

There were, then, two Government cultivations in the first settlement—the old Government House garden which extended northward from near present-day Bridge Street to Sydney Cove, and the Government Farm comprising Phillip's '9 Acres in Corn' and adjoining cultivations on Botanic Gardens Creek at Farm Cove, leased shortly after Phillip's departure at the end of 1792.

Certainly the former garden was used as an official nursery where vines, fruit trees, ornamental plants and other useful exotics were established and acclimatized before perhaps being distributed to the more promising gardens elsewhere, depending upon the supply and the Governor's needs. In 1804 for example, 'Settlers and other Cultivators who may have moist situations on their Farms' were invited to apply for 'that necessary Plant the Bamboo ... from the Government Gardener at Sydney'[32] and in 1806, people 'desirous of cultivating ... valuable timber' were invited to apply for some of the 'very fine Acorns ... saved from the Oaks in the Government Garden at Sydney'.[33] In both instances, the gardener, probably Thomas Alford, promised guidance on planting.

No doubt this Government House garden was also used as a nursery and clearing-house where indigenous plants destined for Sir Joseph Banks, the Kew Gardens, and the gardens of English noblemen and of nurserymen could be tended pending the arrival of a ship boasting of one of Banks's 'Plant Cabbins', fitted with tubs and with tallow candles for heating when necessary. Such a reciprocal arrangement for the exchange of plants was suggested as a practical proposition by Governor King in 1803.[34]

When François Péron, naturalist with Nicholas Baudin's expedition, visited Sydney in 1802, during Governor King's time, he found at Government House

... a fine garden, that descends to the sea-shore: already in this garden may be seen, the Norfolk Island pine, the superb Columbia, growing by the side of the bamboo of Asia: farther on is the Portugal orange, and Canary fig, ripening beneath the shade of the French apple-tree: the cherry, peach, pear, and apricot, are interspersed amongst the Banksia, Metrosideros, Correa, Melaluca [*sic*], Casuarina, Eucalyptus, and a great number of other indigenous trees.[35]

This Norfolk Island pine, *Araucaria heterophylla*, appears in paintings of early Sydney as a landmark of the Government House garden. The generic name *Metrosideros* was then applied to such Myrtaceous genera as *Angophora*, *Callistemon*, *Eugenia* and to some species of *Eucalyptus*. It is interesting to note that this garden was not then entirely cleared of native trees.

Although botanical work with scientific implications was carried out in Government House garden, it seems clear that the idea of an extensive Government Domain surrounding the Governor's residence arose largely from the desire of the early governors to maintain a buffer of privacy in the midst of a penal colony.

It is probable that between 1789 (when agricultural attention turned to Rose Hill) and 1794 (when Nicholas Divine, superintendent of convicts, took up a fourteen-year lease in the Government Farm area) the old cultivations at Farm Cove and along Botanic Gardens Creek were also used as a clearing-house for the import of exotics and the export of indigenous plants. Once the private use of land in this area was curtailed, the cultivations of the former Government Farm were apparently used for such purposes, and despite some inconsistencies in alignments, it is to the furrow lines of the early cereal sowings of those '9 Acres in Corn' that the long rectangular shapes of the beds of the later Middle Garden of the Botanic Gardens have been attributed.[36]

Phillip had had the vision of a vast Crown reserve; Hunter and King had permitted the old Government Farm to be occupied; Bligh had adopted an apparently ruthless policy to restore something of what Phillip had intended. Visionary Arthur Phillip had marked a line, forthright William Bligh had a ditch dug along it—then in 1810, after Bligh had been overthrown, practical Lachlan Macquarie began his sandstone walls. The new Governor made his policy and intentions very clear. His 'demesne' would be inviolate.

Chapter 3

THE GOVERNOR'S DEMESNE 1810–21

I would here beg leave to solicit your Lordship's attention to the importance of the establishment of the botanic garden at Sydney, that has hitherto been attached to the governor's garden at that place, and has derived assistance from the labour of the convicts assigned for its cultivation ... The value of such an establishment, both in affording means of collection and of experiment, and more particularly of diffusing throughout the colony the most valuable specimens of foreign grasses, plants, and trees, is unquestionable ...

COMMISSIONER JOHN THOMAS BIGGE, 1823[1]

Lieutenant-Colonel Lachlan Macquarie arrived in Port Jackson with his 73rd Regiment at the end of 1809, and on New Year's Day 1810 was sworn in as Governor of New South Wales, with the task of restoring the King's authority and peace after the 'revolutionary' period following Governor Bligh's deposition. After an uncomfortable time when the colony had two governors residing near the Tank Stream, Macquarie at last saw ex-Governor William Bligh and ex-Lieutenant-Governor William Paterson, both amateur naturalists, both Fellows of the Royal Society and both Banksian protégés, sail for England in May with the notorious NSW Corps in the *Dromedary*. Paterson died on the voyage home.

Macquarie now had the colony to himself. Reinstatements were made, legal government was resumed and the Governor's own regiment was on hand should due order be threatened. Macquarie took stock of his new situation, and later recalled that he was not favourably impressed:

I found the Colony barely emerging from infantile imbecility, and suffering from various privations and disabilities; the Country impenetrable beyond 40 miles from Sydney; Agriculture in a yet languishing state; commerce in its early dawn; Revenue unknown; threatened by famine; distracted by faction; the public buildings in a state of dilapidation and mouldering to decay ...[2]

Macquarie and his 'dear Elizabeth', like their predecessors, moved into the old Government House which stood near the present intersection of Phillip and Bridge Streets. Government and General Orders soon issued forth. The new Viceroy was quick to note encroachment upon 'the Governor's Demesne'. It seemed strange indeed that Mr Commissary John Palmer, for example, should have two flour mills and a bakehouse within the prohibited confines, and there were other irregularities to be corrected.

If people wished to take a recreational walk, the Governor felt that they could go to that

> ... open Ground yet unoccupied ... hitherto and alternately called by the Names of 'The Common', 'Exercising Ground', 'Cricket Ground', and 'Race Course', bounded by the Government Domain on the North; the Town of Sydney on the West; the Brickfields on the South; and Mr Palmer's Premises on the East, being intended in future for the Recreation and Amusement of the inhabitants of the Town, and as a Field of Exercise for the Troops.[3]

Macquarie 'thought proper to name the Ground thus described "Hyde Park" ...' And it was so.

In August 1810 Macquarie, acting 'on behalf of Government' entered into agreements with Thomas Boulton, stonemason of Sydney, to have two stone walls constructed. One, to be 950 feet long and 8 feet high, 'exclusive of the Foundation', was to be of 'good tough Ashlar on one Side and good rough work on the other side Coped with wrought stone'. The other was a modest 'low wall along the Side of the Government Gardens' 2 feet high and 467 feet long. Four months were allowed for the former; one month for the latter, with payment for the larger work in spirits—460 gallons valued at 30 shillings per gallon.[4] The Governor's celebrated 'Rum Hospital' in the street named after him was to become much better known than his 'rum walls', although the Domain wall was to prove very significant during the last stage of his administration.

In his inimitable way, the late Malcolm Ellis considered the matter of Macquarie's walls:

> Walls were an institution in Botany Bay. No new building's advantages ever were canvassed without reference to its fine, high wall. That round the new military barracks was ten feet high, though the Hospital wall was only nine feet in recognition of the feebler propensity of invalids as climbers. The little orphan girls had a wall round their haven. The Dockyard wall was twelve feet high and the Lumberyard wall had had to be raised to the same level to prevent kindly workers inside from handing out their tools and materials to friends. The Jail paid a tribute to the vigour of the local criminal with a fourteen-foot barrier—and a 'strong' one at that. And even the Government Domain, part of which Macquarie, and particularly his wife, visioned as the flowering garden into which it grew from their none-too-modest beginnings, was protected on the southern side, remote from the main gates and the guard, by ten feet of stone, 'dividing it from the town across the neck of land between Sydney Cove and Woolloomooloo Bay.'[5]

Macquarie was certainly a fervent advocate of gubernatorial privacy and of clear demarcation of boundaries. His enclosure policy provided not only for a stone wall to divide the 'Government Domain' from the town, but also for considerable lengths of post-and-rail fencing, and even for 'pallisading' to be erected by Sydney carpenter John Mould, on top of the existing 'dwarf wall round the Government Domain'![6] Macquarie's report on public works carried out during his administration teems with references to stone walls, brick walls, rail fences, paling fences, stockades and undefined 'strong fences'.[7]

Major-General Lachlan Macquarie, Governor of NSW 1810–21. He strenuously defended the 'Governor's Demesne' as an inviolable Crown reserve.
(Mitchell Library, Sydney)

Having declared in September 1811 that Garden Island was to be 'considered in future as forming a part of the Government Domain',[8] the Governor set about devising rules for the proper regulation of the area he had taken such pains to reserve for the Crown. On 17 October 1812 the residents of Sydney Town were advised:

The whole of the Government Domain at Sydney, extending from Sydney Cove to Woolloomooloo Bay, being now completely enclosed by Stone Walls and Palings, except that part of it which is AT PRESENT UNDER LEASE to Mr Palmer and Mr Riley, where their Windmills and Bakery are erected. Notice is hereby given, that no Cattle of any Description whatever are, after the present Date, to be permitted to graze or feed on the said Domain, those belonging to Government only excepted; and any Horses, Cows, Sheep, Asses, Pigs, or Goats which may after this Notification be found Trespassing thereon, will be taken up and impounded for Damages.

The Public are further to take Notice, that for the future no Stones are to be quarried, or Loam dug within the said Domain, for other than Government Purposes, unless by special Permission obtained from His Excellency the GOVERNOR; and none of the Wood or Shrubs growing within the said Domain are on any Account whatever to be cut down, or otherwise destroyed, on Pain of Prosecution for Felony; and no Boat, except those belonging to Government, are to land in Farm Cove, or on any other Part of the Shore bounding the Domain, except at Bennelong's Point, on Pain of being forfeited, and the Persons who shall have so brought them there severely punished, for Trespassing on the Government Domain, after being thus cautioned by Public Notice.[9]

Such a 'Public Notice' appearing over the signature of the Governor's secretary, John Thomas Campbell, hardly permitted any misinterpretation, although forceful reminders were later proclaimed. The capitalized reference to Mr Palmer's lease was not without significance. This encroachment on the Domain, however praiseworthy in terms of colonial enterprise, had to be removed. Early in November 1813 Macquarie advised John Palmer's creditors that he had 'very lately' discovered that Mr Palmer's lease of three acres and thirty-two rods granted by Governor King 'within the boundaries of the Government Domain (as laid out by Govr Phillip and approved and confirmed by His Majesty's Ministers) ...' had expired on 31 March 1807, after the statutory five years. Mr Palmer or his creditors had until 31 March 1815 to remove the two windmills and bakehouse, for then the Governor would take possession of their sites. This resumption was due to the land 'being required for the very necessary purpose of enabling His Excellency to enclose the whole Domain exclusively within a Stone Wall'.[10] The dutiful Mr Campbell relayed this disturbing decision to Palmer himself in June 1814.[11]

From Macquarie's point of view this little industrial complex was altogether too close, being situated 'on an Eminence overlooking the Government House, and only at a Distance of 160 paces from it'. As the Governor pointed out to Earl Bathurst:

Your Lordship will be at no Loss in Appreciating the Nature and Degree of Annoyance that these Mills and Buildings ... must be to Me and My family, Who, instead of enjoying a Seclusion from the public Gaze, are thus perpetually exposed to it, Wherever We turn thro' the Domain; Even the Passage to the Govern't Garden, which would be a pleasant Place of Recreation to Mrs Macquarie and Myself, lead thro' the Leased Ground and Close by the Mills, Whereby We are Nearly Excluded from that Satisfaction.[12]

Resumption was made, the bakehouse and one (only) of the windmills were demolished, compensation was paid, and the occupants of Government House were presumably thereafter spared both the racket of grinding mill-stones and the aroma of freshly baked bread. In December 1817 the foundation stone of the new Government Stables was laid on the site, where rose Francis Greenway's magnificent castellated building which became the Conservatorium of Music. Like the mills of God, the mills of the law ground slowly, for in 1839 the matter of resumption and compensation was still the subject of a claim against the Government, long after the mills of John Palmer had ceased to turn, and six years after Palmer himself had been laid to rest in St John's cemetery at Parramatta. In a voluminous report with references dating from the laws of Henry IV, Governor Gipps surprisingly declared that he was 'entirely at a loss to find words wherewith ... to express my opinion of such a claim ...'.[13]

Before the security and peace of the 'Governor's Demesne' were finally achieved, there had been other problems to resolve. For example, Nicholas Divine, 'formerly Superintendent of Convicts' who being 'old and infirm, and very deaf' was removed from office and permitted to return to England 'as one of Governor Bligh's evidences', had, as mentioned earlier, long occupied the Government Farm area and an 'Allotment of Ground with ... Houses thereon situated near the Spring and Tanks in this Town'. Divine was still 'in England or elsewhere', but these holdings were to be evacuated and surrendered to the Government by 1 August 1812.[14]

Some people probably felt that the days of Governor Bligh had returned, but once the 'Governor's Demesne' had been cleared and secured, the way was open for the revival of horticultural and botanical activity at the old Government Farm. In fact, Macquarie later listed as one of his achievements 'A Government Garden made on Farm Cove, consisting of five acres of Ground, enclosed partly by a High Paling, with a Brick House for the accommodation of the Chief Gardener'.[15]

By the end of June 1813 twenty convicts were employed on the Governor's horticultural and landscaping schemes for Sydney. In the 'Government Gardens', overseers Nicholas Flood and Thomas Alford supervised the work of Domenico Papillio, Samuel Dent, Sampson Phillips, William Walker, John Agland and Patrick Cooney, while in the 'Government Domain and Park', Nicholas Delaney had oversight of a gang of eleven—Thomas Rider, Thomas Ellis, William Henessey, William Kennedy, John Sly, John Kelly, George Hewlett, John Burke, James Sheridan, James Gill and Robert Francis. Their names are worthy of record,[16] as is the fact that by 1813, there was a recognized distinction between the 'Government Gardens' and 'Government Domain and Park' which contained them.

Macquarie by no means restricted his attention and interest to the environs of Government House, his 'Demesne' and Hyde Park. He promoted exploration, by example as well as by instruction, encouraged road building and instigated a vast programme of public works. In all of these activities, the convict labour force featured largely, and in general, productively.

Before the end of his first year in the colony, Macquarie had gained much first-hand knowledge about the country beyond Sydney. In December 1810 it was proclaimed: 'His Excellency has been much gratified by the natural Fertility and Beauty of the Country in general'.[17] The Governor's company on his explorations of the Nepean and Warragamba Rivers included the pastoralist Gregory Blaxland, whose family was well known to Sir Joseph Banks. Blaxland made two excursions up the Warragamba in 1810, and decided to follow the watershed instead of the gullies to find a way through the Blue Mountains barrier. He recorded that the Governor 'landed to pick some currywing flowers' (probably Kurrajong, *Hibiscus heterophyllus*) at the point of land between the Nepean and Warragamba.[18]

On 11 May 1813 Gregory Blaxland, Lieutenant William Lawson and young William Charles Wentworth, with four servants, set out from Blaxland's farm at South Creek to make an assault on the mountain barrier. On 31 May they reached Mount Blaxland, their westernmost point, and by 6 June had returned to South Creek. Blaxland claimed that he had told Macquarie of his intention to follow the high sandstone ridges between the Grose and Warragamba Rivers, and the Governor had 'thought it reasonable, and expressed a wish that I should make the attempt'.[19] Actually, Macquarie had little time for the Blaxlands, and subsequently tended to attribute the discovery of a way to the western interior to Surveyor George Evans. Finally, however, the Governor publicly acknowledged the 'extraordinary patience and much fatigue' of Blaxland, Lawson and Wentworth in effecting 'the first passage over the most rugged and difficult part of the Blue Mountains'.[20] Nevertheless, as late as November 1816, Blaxland was still pointing out to Banks that the Governor had offered no assistance and little credit for the exploit.[21]

Macquarie promptly despatched Evans to go over and beyond Blaxland's route. Evans left the Nepean River on 19 November 1813 and reached Mount Blaxland a week later. He discovered and traced the Fish River to the confluence of the Campbell and the Macquarie, which he followed for some twenty miles beyond the later site of Bathurst. By 8 January 1814 Evans was back at the Nepean.[22]

During this journey, Evans recorded the ecology of the country in general terms, noted some plants by name and collected some 'samples', including the first timber specimens west of the range, but his chief botanical concern was for 'good grass'. The hot, dry summer of 1812–13 was followed by a dry winter and spring, and by another dry summer, 1813–14. Evans hoped that his discoveries west of the mountains might help resolve increasing problems of feed and overstocking within the County of Cumberland.

By January 1815 William Cox of Clarendon and his small industrious gang of convicts had prepared a road to Bathurst Plains more or less over Evans's route. In April, Macquarie, accompanied by his wife Elizabeth, Surveyor-General John Oxley, Deputy-Surveyor-General James Meehan, Surveyor George Evans, William Cox, Mr Secretary Campbell, John William Lewin ('Painter & Naturalist'), and others, travelled the new road. Six days after Macquarie had proclaimed the town of Bathurst on 7 May 1815, Evans headed south and west to discover and name the Lachlan River.

Macquarie carefully transmitted Evans's journals, descriptions and specimens to Earl Bathurst, Secretary of State for the Colonies. While His Lordship perused this material 'with considerable Interest' he felt that 'all that has hitherto been ascertained of that Country only makes it desirable to penetrate further into the Interior'. The new town would be an advanced depot for achieving this. Although Mr Evans was to be praised, 'he does not appear from the Style of his Journal to be qualified by his Education for the task of giving the Information respecting this New Country, which it is so desirable to obtain'. Earl Bathurst therefore hoped 'that in the further prosecution of these discoveries you will associate with him some person of more scientific Observation and of more General Knowledge'. If the Governor could find suitable scientific personnel, he could consider himself 'authorized to direct another Expedition into the Interior'. If not, then surely 'little difficulty can exist in obtaining scientific persons here, not only willing but anxious to enter upon the great field which has been opened to them'.[23]

To provide future expeditions with adequate guidance, Bathurst provided Macquarie with a detailed memorandum to serve as 'the Groundwork of any Instruction which you may give to future Travellers'.[24] There was the need for 'a detailed Journal' which would record 'all Observations and occurrences of every kind' and descriptions of 'the general Appearance of the Country, its Surface, Soil, Animals, Vegetables and Minerals' together with anthropological data. Care should be taken to look for plants likely to have any economic significance, especially in the fields of medicine, dyeing, furniture making, and shipbuilding. 'Specimens of the most remarkable' plants, animals and minerals should be preserved, and seeds collected 'of any plants not hitherto known'. Thus His Majesty's Secretary of State for the Colonies and former President of the Board of Trade set the pattern for instructions to official exploring expeditions for the next fifty years. Macquarie sent an 'Attested Copy' with his own instructions to Surveyor-General John Oxley.

The year 1816, between the building of William Cox's road and John Oxley's first

*Henry, Third Earl Bathurst (1762–1834). As Secretary of State for the Colonies he sent Governor Macquarie
a detailed memorandum which set a pattern for instructions to official exploration parties.
(from a portrait by Sir Thomas Lawrence in the collection of H.M. the Queen, reproduced by
kind permission of the Royal Library, Windsor)*

expedition to explore the western rivers, was botanically the most significant in Macquarie's term of office. It is also the year from which the Royal Botanic Gardens in Sydney traces its official existence.

Having duly warned the populace of the sanctity of the Government Domain, and revived horticultural activity on the old Government Farm, Macquarie set about in 1813 to implement his wife's scheme for a road to skirt Farm Cove and form a loop at Anson's Point. A section is still known as Mrs Macquarie's Road, and is, indeed, the postal address of the Royal Botanic Gardens. The carriageway followed the northern side of one of the Governor's fine stone walls, part of which still stands in the Gardens, while the inscription in the sandstone rockface at 'Mrs Macquarie's Chair' at Anson's (now Mrs Macquarie's) Point recorded that 'the Road Round the inside of the Government Domain ... Measuring 3 Miles and 377 Yards' was 'finally Completed' according to Mrs Macquarie's plan on 13 June 1816. Today, the Yurong and Victoria Lodge Gates remind visitors that they are passing from the Botanic Gardens to the Outer Domain or vice versa. In Macquarie's time, the 'Government Gardens' were very much part of the 'Governor's Demesne'.

Part of Mrs Macquarie's Drive in 1889. Although probably planted about the time the Drive was opened, many of the Swamp Mahogany trees, Eucalyptus robusta, continue to thrive.
(Royal Botanic Gardens Library)

Three weeks after the opening of Mrs Macquarie's Drive, the Governor considered it was time to issue another warning:

Notwithstanding the repeated and positive Orders which the Governor deemed it expedient to issue and publish, with a View to caution and prohibit Persons from Trespassing on the Government Domain, which has been much injured, not only by Persons breaking down the Wall that incloses [sic] it, but by their cutting down or burning the Shrubbery, destroying the young Plantation of Trees, quarrying of Stones, removing Loam, and stealing the Paling within the said Domain; and notwithstanding the Punishments that have been necessarily inflicted on some idle and profligate Persons ... His Excellency has observed with Regret that Trespasses ... are still continued ...

However, not all were beyond redemption, for

The Orders ... are not meant to extend to the prohibiting the respectable Class of Inhabitants from resorting to the Government Domain as heretofore, for innocent

24

John Paine's photograph of Mrs Macquarie's Chair at Mrs Macquarie's (or Anson's) Point about 1880.
(Historic Photograph Collection, Macleay Museum, University of Sydney)

Recreation, during the Day Time; the Road some Time since constructed round Bennelong's Point furnishing easy Access in that Quarter, and the Gate and Style, at the East End of Bent street, offering free Admission in that Direction.[25]

No guidance was offered concerning the nature of respectability—the inhabitants of Sydney Town knew well enough whether they were qualified to proceed through the Governor's sandstone walls.

The date of completion of Mrs Macquarie's Road is traditionally observed as the foundation day of Sydney Botanic Gardens. Although Macquarie noted that at 1 p.m. on 13 June 1816 'Nicholas Delaney the Overseer of the Working Gang employed for some time past in the Government Domain reported to me that Mrs Macquarie's New Road' was 'completely finished', thereby making this a 'particular and *auspicious Day*', he made no mention of any formal proclamation of the Botanic Gardens, nor it seems, was such a proclamation publicly made. Mr Delaney and his gang of ten were provided with five gallons of spirits with which to celebrate their achievement.[26]

The question of the foundation date is further vexed by evidence that presentations were made to the Governor's Garden (or proposed Garden) before 1816. In 1814, for example, Dr D'Arcy Wentworth's gift of a Norfolk Island pine, *Araucaria heterophylla*, was planted at the intended entrance to the new Government Garden at Farm Cove,[27] and other specimens of the same species had been planted somewhere in the Domain two years before by Nicholas Flood, the overseer.[28] According to a family tradition, still more pines were planted by Major Henry Antill, the Governor's aide, and probably by Macquarie himself, 'when the gardens were commenced'.[29] One of these was apparently the later celebrated Wishing Tree set at the intersection of two major pathways, a little to the east of Botanic Gardens Creek, near the present Visitor Centre.

This became a giant tree which fascinated children until decay led to its removal in 1945. A sculpture presented by Leo and Ida Buring marks this site hallowed by the wishing rituals of thousands. A replacement tree planted on the other (northern) side of Macquarie's wall in 1935 has not yet been accorded the same reverence as its mighty predecessor.

Joseph Henry Maiden, Director of the Gardens, 1896–1924, who, incidentally was quite adamant that 13 June 1816 was 'the official birthday of the Botanic Gardens', discovered several stories about the planting of the Wishing Tree, which by 1906 was 'about 100 ft. high'. It was variously claimed that it had been planted in 1817 by Major Antill (or rather replanted from the immediate vicinity of Government House) at Mrs Macquarie's behest; by Charles Fraser, superintendent of the Garden; by Ned Shakeley, a convict later employed at John Baptist's Surry Hills Nursery; by John Higgerson, ranger of National Park (died 1905); by John Richardson in Mrs Macquarie's presence; by Mrs Macquarie herself. Despairing of the conflicting legends, Maiden boldly engaged in a little playful historical fantasy:

> … I propose to adjust these claims in the following manner:—Ned Shakeley dug the hole, Johnny Higgerson handed him his spade and helped him generally, Mr Fraser turned the plant out of the pot to see that it was all right, and Major Antill planted it with due ceremony. Then Mr Fraser trod the earth about it, staked it, watered it, and tended it during its early days.[30]

In any case, Maiden considered the Wishing Tree to have been 'the most historically interesting plant in our beautiful Garden'.

The mention of Charles Fraser is another reminder of the importance of the year 1816. On 8 April 1816, the transport, *Guildford*, 521 tons, arrived in Port Jackson with 220 male convicts and a detachment of the 56th Regiment including Private Charles Fraser. A native of Blair-Atholl, Perthshire, Fraser had completed a term in the East Indies, and prior to his enlistment at the age of twenty-four, had been a gardener in the service, it was later believed, of the Duke of Norfolk. In November 1816 he was transferred to the 46th Regiment, and in August 1817 to the 48th.[31] It is not known whether Macquarie learned indirectly about Fraser's propensity for botany and horti-culture or Fraser offered his services for developing the revived garden at Farm Cove, but clearly, despite his regimental transfers, Fraser must have performed little mili-tary duty after his arrival. Perhaps the Governor unwittingly anticipated the advice which Earl Bathurst was to pen ten days after the *Guildford's* arrival, and enquired about any scientific interests of members of the new detachment.

The Norfolk Island Pine, Araucaria heterophylla, *long known as the 'Wishing Tree', photographed by
Charles Kerry about 1890.*
(Mitchell Library, Sydney)

However the contact was made, the arrival of one whom John Oxley was to call a
'botanical soldier' just six months before the receipt of Bathurst's scientific memorandum must later have seemed providential to Macquarie. More cause for gratitude
was to come, for in March 1817 a further despatch from Earl Bathurst, dated 1
October 1816, relayed the Prince Regent's request to send 'Seeds of the Choicest
Plants ... of New South Wales' for the Emperor of Austria, with care to be taken to
preserve the seeds from injury by insects or dampness.[32] Then, just before Christmas
1816 King George III's own botanical representative arrived as well—Banks's
appointee, Allan Cunningham.

Having just completed two years in Brazil collecting 'plants &c. for His Majesty's

Botanic Garden at Kew', Cunningham was eager to continue this work in New South Wales. No doubt he was still conscious of the humbling terms in which Banks had advised him of the appointment:

You have had the good fortune to be selected[33] from among the very great number of Excellent young men who have been Educated at the Royal Gardens of Kew under the Eye of their worthy director Mr. Aiton, not so much from any superiority you possess of many others in Botany or Horticulture as from a firm persuasion in Mr. Aiton that you do Excell in [the qualities] of honesty sobriety diligence activity Humility & Civility & that you will never lose sight for a moment of these Essential Qualifications which above all others insure to a traveller Respect among Strangers & assistance from those in high office who have the Power either of giving or of withholding it.[34]

Cunningham was to 'use Every possible opportunity of sending home pacquets of seeds carefully sealed up & adressed to Wm. Aiton Esqr under cover to the Earl of Liverpool & you will keep a daily Journal of your Proceedings noting every thing proper to be made made [*sic*] known'.[35] Expenses were to be met by drawing bills upon Banks, taking care 'to procure the most advantageous Rate of Exchange'. The actual collecting process was to be a rather exclusive business requiring great care:

You are not ... to allow any Person whatever to Receive or under any Pretence to Obtain from you any Part of the seeds or any of the Plants or bulbs Collected by you while you Continue in your Present Employ Should any new Plant sent ... by you to Kew appear in any other Garden an Enquiry will be immediately set on Foot to Find out in what way ... it was procured & if ... it Proves to have been obtained from you in any Circuitous manner whatever your having Parted with ... it will be deemed a breach of the Fidelity you owe to your Employers.[36]

It was Banks's final instruction that led to strife between the King's Representative and the King's Botanist. Cunningham was not only to transmit to Banks his journals, but also

... to write to me as often as possible Stating to me such Circumstances of your Reception & the Conduct of those with whom you are Concerned Towards you & your undertaking as you may think fit to Communicate but you are Carefully to abstain from all Enquiry Relative to the Political State of the Colony.[37]

With these instructions, a salary of £180 per annum, and a request to be frugal, Allan Cunningham, King's Botanist, twenty-five years old, waited upon His Excellency the Governor of New South Wales at his country residence at Parramatta on 21 December 1816.[38]

Despite the receipt of a despatch from Earl Bathurst in June 1815 advising Cunningham's appointment, Macquarie observed 'that he had received no Instructions' concerning Cunningham but he received the botanist 'with every possible degree of kindness and hospitality'. Macquarie 'strongly recommended' that he should join Charles Fraser on Oxley's forthcoming expedition to trace the course of the Lachlan River, being 'firmly persuaded that an Immense number of new and interesting specimens of Plants might be detected, in the several districts thro' which we should pass'.[39]

Cunningham was an enthusiast. Having settled in a cottage at Parramatta, he familiarized himself with the flora of the Sydney-Parramatta area and generally prepared himself for the botanical work ahead. As he had pointed out to Banks in a lengthy postscript to his original application for employment:

It is a love of plants, and to search for them in their wild state, and a wish to make myself useful in the Capacity of a Collector, that now urges me to address you at this time, and should I be so fortunate as to be appointed Collector in the Service of His Majesty's garden, it shall be the highest ambition of my life to exert myself in the perform[ance] of the requisite Duties that constitute a Collector, so that the Royal Collection at Kew may exceed all other Collections in the Riches of new, beautiful and desirable plants.[40]

This enthusiasm, his status as King's Botanist, his powerful patron and possession of what amounted to a royal commission to perform his duties, all combined to bring about the impending clash with Macquarie.

Macquarie was confident that a man of Oxley's 'Talents and Qualifications', with George Evans as second in command, to be accompanied now by two botanists and 'William Parr, mineralogist' would surely produce evidence of a scientific examination of the interior to Earl Bathurst's satisfaction. Although the Governor had strongly recommended that Cunningham should 'embrace so very favourable an opportunity' by joining this expedition, the young botanist still had doubts almost to the point of departure. He asked Banks for further instructions, but took the Governor's advice, and on 20 March 1817 'sent forward a specimen press and some paper' to the Bathurst depot.

Clearly Macquarie had decided by the end of 1816 to send Fraser with Oxley's expedition. On 22 February 1817 the Governor drafted his 'Instructions for Private Charles Fraser of the 46th Regt of Foot, Botanist and Gardener' directing Fraser to place himself under Oxley's orders. Notwithstanding his specialized duties, the botanist was not to 'consider any Duty required ... such as Carrying Provisions, leading the Pack Horses, or rowing in Boats—as incompatible with the more immediate Duties' of making

... a separate Collection of Botanical Seeds and Plants ... for His Majesty's Ministers at Home and for this Government, (independent of that to be made by Mr Cunningham the King's Botanist ...) You are accordingly to make a Collection *in Triplicate* of all rare and new Plants and seeds you meet with ... packing up and preserving the same with all possible care, so as to deliver them all safe here to me on the Return of the Expedition ... You are on no account to give away any of those Plants or Seeds ... to any other Person whatsoever excepting to me.—But you are at liberty to give any reasonable assistance in your power to Mr Cunningham.[41]

A month later, Macquarie referred to Fraser as 'colonial botanist', thereby distinguishing him from 'Mr Allan Cunningham, King's botanist'.[42] Cunningham referred to his botanical colleague as 'collector for Lord Bathurst'.[43]

By the time Cunningham arrived at Bathurst, Fraser had been botanizing in the surrounding hills for about a month, and quite early in the expedition Oxley advised the Governor: 'I think Fraser will be enabled to present to your Excellency a Valuable Collection. I cannot say as much for the mineralogical acquisitions ...'[44]

The journals of Oxley and Cunningham recorded that the two botanists worked co-operatively, and their excursions added greatly to the area of country traversed. Cunningham however, seldom referred to Fraser by name, but often noted plants 'we' had discovered. When Cunningham was 'seized with a violent ague' and unable to stir, Evans, Fraser and Parr brought plants to him.[45] Cunningham had taken 'moderate-sized portable saddle bags, and specimen cases well canvassed over and painted, for the reception and protection of those treasures that the interior of this country may afford'[46] and doubtless Fraser was similarly equipped. Cunningham also brought 'a quantity of peach-stones of two qualities, some quince pips or seeds, and a few acorns'[47] which he sowed throughout the journey.

On the return to Bathurst, Oxley sent the Governor a report in which the botanists were warmly praised:

It would perhaps appear presuming in me to hazard an Opinion upon the Merits of Persons engaged in a Pursuit of which I have little knowledge. The extensive and Valuable Collections of Plants formed by Mr A. Cunningham, the King's botanist, and Mr C. Frazer, the Colonial Botanist, will best evince to your Excellency the Unwearied Industry and Zeal bestowed in the Collection and Preservation of them. In every other respect they also Merit the highest praise.

Macquarie duly relayed this pleasing information to Earl Bathurst.[48]

Cunningham was loath to let the results of his 'Unwearied Industry and Zeal' out of his sight: 'Rather than subject my luggage to accident in passing rivulets, I determined to accompany the whole myself, giving up my saddle horse to bear that part of my collection that could not be carried by the cart'.[49] This collection was enlarged on the way back over the Blue Mountains, and indeed Cunningham added to it until he 'arrived at Parramatta at dusk with the whole ... collection' on 8 September 1817.[50]

The next morning Cunningham reported to the Governor who suggested that he should now join Lieutenant Phillip Parker King's coastal survey expedition. That evening, and again on the 17th, the botanist dined at Government House. On the 18th, however, the Superintendent of Government Stock demanded the return of the horse Cunningham had used on Oxley's expedition.[51] Hoping to retain the horse for local excursions and for the transport of specimens, Cunningham wrote to the Governor accordingly, but this request drew a cool, if immediate response:

... I very much regret that I cannot consistently with the nature of any Instructions from Home comply with the request contained in your letter ... namely furnishing you with a Government Horse to enable you to carry on your Botanical Pursuits in this Country. This is a Govt indulgence tho' applied for by them has been even refused to the Surveyors and Medical Officers of Government, whose various Public Duties frequently require the use of a Horse; and were this indulgence extended to you, they would have reason to complain of so mortifying a distinction.

If a Horse is requisite for enabling you to prosecute more efficiently your Botanical researches I should imagine your Employers at Home would approve of your Purchasing one.[52]

Cunningham contented himself by confiding the gist of this letter to his journal, having 'returned the horse forthwith without delay'. He wrote to Banks and Aiton to advise his safe return and prepared his specimens for shipment by the brig *Harriet*. During October and November he continued his excursions, presumably chiefly on foot, and visited ships likely to bring tidings from Banks.

On 1 December Cunningham again wrote to Banks concerning the recent expedition, the matter of accounts, the business of the 'Government Horse', the unsatisfactory nature of plant cases and his relationship with the Governor. He also confessed

Allan Cunningham, King's Botanist and explorer, who later became Colonial Botanist and Superintendent of the Botanic Gardens, February 1837–January 1838.
(Mitchell Library, Sydney)

that 'several Persons (Prisoners) of the late Expedition' had collected 'ample Dupli-cates of many of my very interesting and valuable Seeds and bulbs ... chiefly with a view of turning them to Cash, upon their return to Sydney'. It was understood that 'many are now in the Possession of several wealthy Individuals ... who intend to transmit them to their friends in England by the earliest opportunity ...' Cunning-ham hoped that 'nothing incorrect' might be attributed to him, in the light of his instructions.

The contents of this letter were clandestinely relayed to the Governor[53] and the next day, 2 December, Cunningham 'waited upon His Excellency, according to appointment, in order to superintend the execution of a few drawings of plants dis-covered in the interior, which the Governor intends to transmit to Earl Bathurst'.[54]

Macquarie took great care in reporting to Bathurst the botanical success of 'the late Circuitous Tour thro' the Interior'. It had 'in its Botanical Department been produc-tive of an Accession of upwards of Five Hundred Plants totally different from those hitherto Collected or known in this Country. One Hundred and Fifty of them were found bearing Seed.' The Governor further advised Bathurst that he was sending 'Dried Specimens of all the Plants'; seeds of 150 species, and drawings 'by the Master-ly Hand of Mr Lewin' of four of the plants which

... were Considered so rich and beautiful by the persons, who Collected them ... These Drawings being taken whilst the Plants retained some Share of their Natural

Beauty, and immediately under the Eye and Direction of the Botanists who collected them, their Colours and Peculiarities have been well preserved, and will Convey a much more perfect Idea of the Plants themselves than Could be possibly Obtained from the bare Inspection of the dried Specimens, especially after so long a voyage as that they are about to Undergo.[55]

In addition, there were packets of seeds for the Emperor of Austria and for 'Monsieur Goüm, Superintendent of the King's Garden at Paris, ... particularly addressed to ... Sir Joseph Banks', and there were 'thirteen Amaryllis Bulbs and twelve Bulbs of the Pancraticum Macquaria'.[56]

On 16 December 1817, having loaded his luggage on the cutter *Mermaid* in preparation for the northern voyage, Cunningham made a courtesy call on the Governor, 'actuated by the purest motives'. What followed is best described by the botanist himself in a letter to Banks:

The Governor ... wish'd to know whether I was satisfied with the Assistance he had render'd me, ... I answer'd that I had hoped to be provided with a small House or Hut, a Government Horse, and any other Indulgence, which from the situation I held ... and more especially from Instructions, which I doubted not, His Excellency had received from the Colonial Office respecting me, I humbly presumed His Excellency would have deem'd it justifiable to afford me; but for the Assistance I have already received, I humbly beg'd His Excellency would accept my sincerest thanks ... He hinted that his Instructions (affecting me) were in the most common & general Terms, and that those Indulgences I did enjoy, were afforded me by His Excellency more from a favourable impression he had received of me ... than from any Commands from home. His Excellency finally concluded, charging me with having written to you (by this ship) against himself, upon this subject, and that he had obtain'd it from good authority.

Cunningham, 'with becoming respect', tried to assure the Governor that he 'had merely detail'd matters of fact' advising Banks of the support he had indeed received, and that he had by no means sent any 'accusation or charge against His Excellency', but the botanist was 'not allow'd to make any further observations' for the Governor abruptly left the room. Cunningham quitted Government House 'under an ill Impression of the sinister means' used to ascertain the contents of his correspondence. It was 'doubtless thro' the medium of a Constable, who had been recommended ... by the Magistrate, as a useful man and who I had employ'd copying my Journal'. Seeing possible grounds for winning favour, 'this Prisoner ... had reported the subject to His Excellency'.

Fearing that his letters may have been 'seized or Detain'd', Cunningham boarded the *Harriet* where he found 'all ... Packets safe'. He explained to Banks that in describing the assistance he had, and had not, received from the Governor, he was in fact obeying Banks's own instructions. No doubt with great relief, Cunningham added, 'I expect to sail with Mr King tomorrow morning'.[57]

Macquarie ensured that Banks heard his side of the story. He held a 'very high Respect' for Banks 'as a Patron of Men of Genius and Science'. It was therefore 'with much Disappointment and Surprize' that he

... learned that Mr Allan Cunningham ... a Collector of Seeds and Plants, and generally ... an Operative Botanist, has addressed a Letter to you ... Complaining that I have not extended those Facilities towards him in making his Collections that He had been led to Expect and ... that I have with held All these Accommodations necessary to his personal Convenience and Comfort.

While reluctant to believe the report, Macquarie felt 'that some little Enquiry should be made'. On 'putting a few Questions' to Cunningham, he soon felt, because of the botanist's 'Confusion and Hesitation of Manner' that the 'Information was Correct'. Such 'false and ungrateful Conduct' had caused Macquarie to express his regret and disappointment, and the botanist was 'dismissed ... with a suitable Reproof'. He stressed that he had been hospitable to Cunningham, who had been received 'with the Civility and Attention due to a Gentleman'; he had 'Complied with all his Demands', and Banks could 'Judge how far I have merited such a Return from this Unbred, Illiterate Man, whose only pretensions to personal Attention from me arose from the Opinion you have entertained of his Usefulness in the Line of his Profession'. Cunningham could hardly have been allowed a house 'whilst several of the principal Officers of this Government' were not so accommodated, but he had been given 'one Government Servant permanently', and 'Rations equal to those I draw for my own use'.[58]

On 22 December 1817 two vessels cleared the heads of Port Jackson. Allan Cunningham was aboard one bound for the south, west and northern coasts and more botanical discovery, while aboard the other were the botanical results of his expedition with Oxley, and some rather unhappy correspondence, bound for England and Sir Joseph Banks.

Banks, now old and gout-ridden, retained his customary diplomacy, without allowing complimentary phrases to obscure his feelings. In July 1818, he acknowledged Macquarie's letter 'Complaining of the Prince Regents Botanist ... for having written to me on the Subject of Certain Indulgences which he expected'. There was, however, 'nothing in Cunningham's letters' which 'had the appearance of a Complaint', a point with which Earl Bathurst had agreed. Banks was pleased that Cunningham had reported 'the indulgences of a Servant & Rations', even if it were necessary to withhold 'the house & the horse' which Cunningham 'would have been glad to have Received'. No doubt 'at the time', His Excellency 'had sufficient Reason' for his conduct. 'If at any future time it should be deem'd expedient' to grant Cunningham both these indulgences, then Macquarie's 'Countenance to scientific Pursuits will be brought into Notice by the diminution of ... Expenses on the annual audits of the Collectors accounts'. Sir Joseph and Earl Bathurst could only suppose that the Governor had 'been Greivously misinformd' [*sic*].

Happily, there was a credit side. Macquarie's 'activity in Causing the Country beyond the blue mountains to be explord' had won 'the approbation & gratitude of all Scientific Persons here ... & we feel Certain that the interesting business of discovery will never cease', while the colony had such a Governor. Could Macquarie make some enquiries concerning 'the Duck billd duodruped'? Does it really lay eggs? Banks would not credit 'this most extraordinary anomaly' and some further information would be most welcome if the Governor could but spare the time.[59]

Banks then turned to Cunningham, whom he congratulated on the 'most valuable Collection of seeds & specimens the Fruits of your Journey beyond the blue mountains'. Cunningham's conduct had 'given Perfect satisfaction & ... ensured ... a Continuation of the Friendship & Countenance of all here who are adicted to the study of Botany'. Macquarie 'very improperly' found fault with Cunningham 'for Complaining of his Treatment', and the Governor had been informed that the botanist had acted 'agreeably to ... instructions'. Banks feared 'there is some jealousy in your Governor in favour of his Colonial Botanist who you do not mention but who has I am informd sent home some Plants different from [those you sent (?)]'.[60] Banks hoped that his letter to the Governor 'will induce him to give you more encouragement than he has done', but even if not, he added significantly, 'you will recollect that he is soon to come home & is likely to be Replaced by a more Scientific Governor'.[61]

Meanwhile, Charles Fraser was busy between the Surveyor-General's expeditions, preparing additional collections for Macquarie to send to Earl Bathurst. Just after Oxley's second expedition set out, Macquarie despatched by the ship *Lady Castlereagh*, 'several Cabins, Tubs and Cases of the most admired rare and choice Flowers, Shrubs and Plants of this Country'. These included 'Gigantic Lillies' (*Doryanthes excelsa*), 'Rock Lillies' (*Dendrobium speciosum*) and 'Norfolk Pine Plants' (*Araucaria heterophylla*), bound for Emperor Francis I of Austria, Prince Leopold (later King of the Belgians), and Queen Charlotte, George III's consort, to whom the Governor had 'ventured to address a Proportion of these curious Productions' with the 'request that your Lordship will have the goodness to Present my Dutiful Respects herewith to Her Majesty'. Special care was taken with the selection for Prince Leopold, as Macquarie had 'learned from Colonel Addenbrooke that His Serene Highness wished to possess some of our most rare and choice Productions in this Kind'.[62] To ensure that the living plants would receive ample care, Macquarie 'Selected a Person named Alexander Colley', a convict 'intelligent as to the manner of treating Plants', and granted him a pardon and 'Free Passage, the better to secure his care and attention'. It was a botanical disaster that after all this, the *Lady Castlereagh* should have been 'Wrecked on the Coast of Madras'.[63]

Allan Cunningham was examining Apsley Strait, between Melville and Bathurst Islands, when Oxley left Sydney on 20 May 1818 to trace the Macquarie River because of 'the most sanguine expectation ... that either a communication with the ocean, or interior navigable waters, would be discovered by following its course'.[64] George Evans again went as second in command, and Macquarie

... directed Charles Frazer the Colonial Botanist, to join ... the Expedition ... for the purpose of making Botanical Collections of all rare Trees, Plants, Shrubs, Flowers, and Seeds of Trees, Plants, &c. as he may discover ... and as he is to be exclusively employed in making these Collections for His Majesty's Ministers at Home—and for the Government,

Oxley was 'to afford him every reasonable Assistance'.

Remembering the problem of private botanical enterprise in the earlier expedition, Macquarie enjoined Oxley 'to prohibit all other Persons Employed on the Expedition from interfering with Charles Frazer in making his Collections, by making similar Collections for Themselves and their Friends, which you are not to permit them to do'.[65] Clearly the Governor was more concerned to please the conscientious Earl Bathurst than the aged confidant of a sick King.

On 26 May 1818, just before the expedition left Bathurst, Fraser expressed gratitude to his viceregal patron:

May it please your Excellancy

To excuse the Liberty I thus take in writing to you

I yesterday discovered a quantity of the Amarylis Australis, which was brought from the Banks of the River Macquarie at Wellington Vale, they are the same as those discovered in 1817 on Macquaries Range, they are a Bulb of great value, and will be very desirable for his Serene Highness the Prince Leopold of Cape Colony [sic], I have sent direction to the Gardener, herewith, relative to the planting of them. If an opportunity serves, I will send some more from Wellington Vale.
The other Seeds and Specimens which I have collected since my departure, I send herewith.
With every mark of respect and Sincere Gratitude, I have the Honor to be

> Your Excellancys
>
> Most Obedient and very Humble servant,
> Charles Fraser.[66]

In contrast to the apparent wilfulness of the King's Botanist, the tractability of the Governor's Botanist must have been as welcome to Macquarie as the bulbs 'of great value' of the Darling, Murray or Macquarie Lily, *Crinum flaccidum*.

At the end of June 1818 the expedition encountered the Macquarie Marshes, which abounded with the Common Reed, *Phragmites australis*, 'six or seven feet above the surface',[67] thus making the course of the main stream virtually impossible to trace. Surrounded by an 'ocean of reeds', Oxley believed in the light of his similar experience on the Lachlan that he was 'in the immediate vicinity of an inland sea, or lake'. Thwarted by the reedy labyrinth, Oxley turned east to cross the newly discovered Castlereagh, the Warrumbungle Mountains and the northern part of the Liverpool Plains before following the Hastings River to Port Macquarie, and the coast to Port Stephens.

Fraser experienced a wide variety of landscape on this journey, ranging from salt-bush plains to sub-tropical rainforest and there were considerable difficulties in transporting material over mountain ranges and across numerous streams. Certainly Oxley was pleased with his work: 'Mr Charles Fraser, the colonial botanist, has added many new species to the already extended catalogue of Australian plants, besides an extensive collection of seeds, &c.; and in the collection, and preservation, he has indefatigably endeavoured to obtain your excellency's approval of his services'.[68]

As a result, in March 1819 Macquarie advised Earl Bathurst that he was sending by the *Shipley* 'three Cases containing all the rare and Choice Plants discovered and Collected by Mr Charles Frazer, the Colonial Botanist, during the ... last Expedition of Discovery under Mr Oxley.' These cases were addressed to Prince Leopold, the Emperor of Austria and to Bathurst himself, and were given into the care of the Governor's former aide, Lieutenant John Watts, 46th Regt. Paintings by John Lewin of eight of Fraser's specimens accompanied the collection, and so did a request from Fraser for 'Certain Books treating on the Science of Botany'[69]—a request which Bathurst and the Lords Commissioners of the Treasury approved. The books were

W.T. Aiton, *Hortus Kewensis*, London, 1810–1813; Robert Brown, *Prodromus Florae Novae Hollandiae*, London, 1810; and C.H. Persoon, *Synopsis Plantarum*, Paris, 1805—solid fare for anyone, 'botanical soldier' or otherwise.[70]

Earl Bathurst expressed his 'sincere acknowledgement' of this new material, adding that the collections addressed to him were assigned 'to the different Public Establishments in this Country, to which he considered that they would be most acceptable'.[71]

In July 1819 Macquarie sent yet another shipment by the *Surry* of 'Australian Seeds and Plants'—the result of further work by Fraser[72]—which was entrusted to Captain Thomas Raine, who would 'have the honor of delivering them at Downing Street'.[73] To compensate for the loss of material sent on the *Lady Castlereagh* in June 1818, the Governor despatched a further collection of 'Plants, Seeds and Geological productions' for the Prince Regent, the Emperor of Austria, Prince Leopold and Earl Bathurst, in the hope that they would 'prove Acceptable to the Royal and Illustrious personages, for Whom I have had the honor to procure them'.[74] These collections were sent aboard HMS *Dromedary* in the charge of '*John Richardson* (now a free man by Absolute Pardon) a Gardener by Profession'.[75]

While Macquarie thus endeavoured to meet the botanical requests of the 'Illustrious personages' at home and on the Continent, there continued to be many colonists anxious to avail themselves of any advantages arising from the work undertaken in the Gardens and Domain by Fraser and his men. Apparently these requests became embarrassingly frequent, for in July 1819 the Governor sought to curb local horticultural enthusiasm:

Numerous Applications having been made by different Persons to the Gardener and Nursery Man, at the Government Garden and Plantation at Farm Cove, for Seeds, Plants, Shrubs, and Trees of various Kinds, it is necessary to inform the Public, that Thomas Currie, Gardener, &c. to HIS EXCELLENCY the GOVERNOR, has received ... Instructions not to attend to or comply with any Orders or Applications of the above Kind, or other than such as HIS EXCELLENCY may himself be pleased to issue or authorise.[76]

This matter of a government institution supplying agricultural and horticultural material to individuals was long to badger successive governors and superintendents.

Meanwhile, Allan Cunningham arrived back in Sydney from his voyage with Lieutenant P.P. King on 29 July 1818, and returned to his house at Parramatta. On 12 August, contrary to the Governor's expectations, he called at Government House, Sydney, to pay his 'humble respects' before bringing correspondence and journal up to date, and preparing specimens for shipment to England. As King was uncertain about the departure time of his next voyage, Cunningham planned an excursion to 'the Five Islands' or Illawarra, and he accordingly asked the Governor 'for a light Government cart, a horse, a spare pack saddle, etc.'[77]—a sure test of the vice-regal temper. Cunningham explored the gorge country near Parramatta while he waited for Macquarie's reply. Failing to receive one after eight days, the botanist called on the Governor at Parramatta, but he 'was stated to be from home'. Cunningham finally received an affirmative reply in person at a muster parade on 12 October.

The month's journey through Liverpool and Appin to Illawarra was highly productive. Cunningham noted by name, and probably collected, at least 150 species, including the celebrated Illawarra Flame Tree, *Brachychiton acerifolius*. Although botanically fascinating, this area presented many difficulties, the most formidable of which were the steep escarpment and the density of the rainforest; there were also fog and rain, ticks and leeches. To cap it all, the 'Govt Horse ... lost all his shoes on ascend^g the steep mountain'. This horse had already baulked at swampy ground, so that Cunningham had to hire another, as he put it, 'the Gov^t one, allow'd me, turning out bad'.[78] One is tempted to wonder just what kind of steed Macquarie had approved for release to the botanist!

Cunningham joined Oxley, King and Fraser in 1819 on a survey expedition to Port Macquarie and the Hastings River, where the botanical harvest was again rich, having much in common with Illawarra. His waterstained folio journal provides a vivid picture of the virgin rainforest of the northern rivers as exemplified by the Hastings just two years before the settlement of Port Macquarie brought convict timber-cutters.

Returning to Port Jackson on 9 December 1820 after his third voyage with Lieutenant King on the *Mermaid*, Cunningham landed yet another botanical cargo only to learn that King George III had died on 20 January and Sir Joseph Banks on 19 June. Now bereft of the support of Kew's royal patron and of his 'excellent and invaluable friend' Banks, and relying on the capricious favours of a Governor who could be decidedly unfriendly, the botanist must have felt depressed and insecure. Nevertheless, Cunningham joined King on two more voyages to the northern coasts, no doubt encouraged by the letter of commendation Banks had written just eight weeks before his death.

By now the Governor was not without his own problems, for opposition to certain aspects of his administration had long been mounting. One such aspect was his almost obsessive protection of the Domain. For example, in April 1816, William Henshall, Daniel Read and William Blake, on being apprehended, gave various reasons for penetrating the Domain's defences, for which they were summarily administered twenty-five lashes each without any magisterial hearing. On 12 June, the day before Mrs Macquarie's Drive was completed, Judge Jeffrey Hart Bent sent the offenders' depositions to Earl Bathurst as clear evidence of the Governor's arbitrary administration of justice.[79]

Exhibiting a curious mixture of moral, judicial and botanical concern, the Governor who had so encouraged the reinstatement of emancipists to society, continued to fulminate against trespassers who flagrantly avoided passing through the Domain gateways he had provided:

... these regular Public Entrances did not suit the Persons going thither for vicious and disorderly purposes, namely secreting stolen Goods, which have been found there frequently, and for which many parts of it are well Calculated, being wild, rocky Shrubbery, which had remained undisturbed by the Hand of Civilized Man. This Shrubbery was also much frequented by lewd, disorderly Men and Women for most indecent improper purposes. I had long wished to put a stop to these disgraceful Meetings and indecent assignations, as well as to save the Shrubbery and young Plantations of Forest Trees, which had been planted in the Grounds ...[80]

While Charles Fraser was assiduously collecting botanical material for the Governor and Earl Bathurst during Oxley's expeditions of 1817–19, Bathurst had decided that an enquiry should be made into the affairs of New South Wales. John Thomas Bigge, former chief justice of Trinidad, was appointed Commissioner of Enquiry, to investigate the settlements of New South Wales 'as fit receptacles for convicts'. The investigation would be penetrating, to the extent that His Majesty's Government anticipated a report bearing upon 'the police, the agriculture, the commerce, the revenue' and 'the state of society in the settlements'.[81] And so on 26 September 1819, the peace of the Governor's Domain and Gardens was disturbed by the booming of a thirteen gun salute as the *John Barry* sailed down the harbour bearing the Commissioner who was to make a searching enquiry into the administration and general state of the Colony. The inquisition was to last seventeen months, and from the enormous mass of evidence, three published reports would emerge.

Charles Fraser was considered sufficiently expert to accompany the Commissioner on excursions into the interior of New South Wales and to Van Diemen's Land in 1820. Bigge thought highly of the botanist, and directed Earl Bathurst's attention to

... the importance of the establishment of the botanic garden at Sydney, that has hitherto been attached to the governor's garden at that place, and has derived assistance from the labour of the convicts assigned for its cultivation. It has been lately placed under the management of Mr C. Frazer ... and by his care and attention the collection has been enriched with all the most curious plants ... discovered in the course of ... expeditions, as well as by contributions from New Zealand, the islands of the South Seas, Bengal, and China.

The Commissioner clearly appreciated the scientific and economic value a botanic garden could have if properly administered:

The value of such an establishment, both in affording means of collection and of experiment, and more particularly of diffusing throughout the colony the most valuable specimens of foreign grasses, plants, and trees, is unquestionable; and I have great satisfaction in stating that, as far as his means have allowed, these benefits have been realized under the zealous exertions of the present colonial botanist.[82]

As part of his enquiry, Bigge had called upon Fraser to supply a catalogue of the plants then growing in the Botanic Garden. The list provided ample evidence of Fraser's own collecting expeditions and of correspondence with other institutions.[83] By 1820 Charles Fraser clearly had the Botanic Garden, as such, well established as an institution in its own right, quite separate from the Governor's kitchen garden.

It is probably significant that shortly before Bigge left the colony, the botanist's status and salary were ratified by a formal appointment: 'His Excellency the Governor has been pleased to appoint Mr Charles Fraser to act as Government Colonial Botanist in New South Wales with a Salary of Five Shillings per Diem, until His Majesty's Pleasure shall be known'. Fraser would be paid from the Colonial Police Fund as from 1 January 1821,[84] and shortly there was formulated a statement of duties which throws an interesting light not only upon the function of the Colonial Botanist but also upon the purpose of a Colonial Botanic Garden as viewed at that time:

The Duties of the Office are principally the culture of such Exotics as are from time to time introduced into the Garden from various parts of the Globe and which are afterwards distributed in considerable portions throughout the Colony and its dependencies. The Collecting and remitting of Indigenous Plants and Seeds to various persons and countries who have contributed to the Collection. The propagating of all Sorts of Fruits and Grasses that are interesting. And the forming of a General Collection and arrangement of Plants.[85]

While the nature of 'His Majesty's Pleasure' was awaited, Fraser was secure as Superintendent of the Garden and Colonial Botanist, although the salary of £91-5-0 per annum was hardly embarrassingly munificent (the Surveyor-General's salary, for example, being £273-15-0). With his own 'Pleasure' doubtless limited, Fraser felt bound by September 1821 to make a bid for some further recognition:

The Humble Petition of Charles Fraser,
Respectfully States

That Memorialist arrived in this Colony in the year 1816, and is now holding the appointment of Colonial Botanist, and acting in that Capacity attended J. Oxley Esqʳ Surveyor General on his expeditions into the interior.

That memorialist has never received any indulgence of Land in this Coloney, and it being his intention to reside herein, respectfully solicits your Excellancy may be pleased to grant him such a portion as he shall be considered deserving of ...

The Governor obligingly endorsed the petition on the day of its presentation, granting '*500 acres of Land*, with *three* Government Men on the Store for six months ...'.[86] As will be noticed later, it was also the day on which Macquarie and Fraser together inspected and reserved a site for an entirely new botanic garden, which, unfortunately, was never developed.[87]

Having thus made fitting final gestures to his Colonial Botanist, Macquarie journeyed to the Hastings and the Hunter Rivers returning on 21 November 1821, when he had first 'the happiness of meeting ... dearest Elizabeth' and then the 'agreeable surprise' of making the acquaintance of his successor strolling in the Domain he had striven so hard to preserve.[88]

Having accepted Macquarie's resignation in July 1820, Earl Bathurst asked Arthur Wellesley, Duke of Wellington, newly appointed to Lord Liverpool's cabinet, to suggest a successor. The Duke nominated Brisbane whom he had known as a soldier and a friend for some thirty years. Accordingly Brisbane again applied for the position in October, pointing out to Bathurst that he understood his wishes had been communicated 'by His Grace the Duke of Wellington and by the late Sir Joseph Banks'. Although Bathurst clearly appreciated the worth of a man who had experienced no less than fourteen battles, twenty-two minor actions and six sieges, who enjoyed a high reputation as a scientist, more especially as an astronomer, and who was a Fellow of the Royal Society and a Knight Commander of the Order of the Bath, he is alleged to have drolly advised Wellington that he 'wanted a man to govern, not the heavens, but the earth'.[89] Nevertheless the appointment was made. On 1 December the faithful Mr Campbell, now provost-marshal, who had issued so many proclamations regarding the Domain and the Gardens, read the new Governor's Commission in Hyde Park. Thus 'the Old Viceroy' was at last replaced by Banks's 'more Scientific Governor', Sir Thomas Brisbane, who had first applied for the position in the year Macquarie had travelled over William Cox's new road to proclaim the settlement of Bathurst.

Chapter 4

A SCIENTIFIC INSTITUTION? 1821–48

… it is presumed, the primary object to be kept in view, in conducting such an establishment at Port Jackson, is to render it valuable to the Colony.

ALLAN CUNNINGHAM, 1832[1]

Between November 1821 and February 1822 the colonists of New South Wales again had the pleasure of the presence of two King's Representatives, one just commissioned, and the other acting almost as if he still were. Happily they remained on cordial terms.

A fortnight after Sir Thomas Brisbane was commissioned, Macquarie accompanied him to Parramatta, and thereafter proceeded on 'a tour of inspection of Bathurst'. Then in January 1822 Macquarie visited the Illawarra area, where he graciously named Mount Brisbane, and noted 'the largest and finest box trees' he had ever seen, and a giant red cedar 'which measured 21 feet in diameter and 120 feet in height'. As he travelled over Cornelius O'Brien's newly constructed pass road down the Illawarra escarpment, 'frightful to look at, but perfectly safe', the ex-Governor may even have spared a thought for Allan Cunningham, King's Botanist, and the unfortunate 'Government Horse' which lost its shoes.[2]

While one Governor was enjoying the scenery of Illawarra, and the other was planning an observatory at Parramatta, Charles Fraser wrote to Frederick Goulburn, Colonial Secretary, suggesting that 'His Excellency the Governor' might approve 'That a Conservatory should be built in the Governors Gardens Sydney … Length, Fifty feet, greatest height, 20 feet, greatest breadth 20 feet,' so that the Colonial Botanist after 'an experience of nearly six years' would at last, with 'the assistance of Glass' be able to cultivate 'many Tropical Fruits, culinary, medicinal, and other Plants, so as to perfect their Seeds'.[3] For the time being, however, it seems that tropical imports were to endure the uncertainties of Sydney's weather with rather more improvised protection.

Cunningham returned to Port Jackson with Lieutenant P.P. King in the *Bathurst* on 25 April 1822, to find that the *Surry*, which had brought him to the colony at the end of 1816, had recently conveyed Macquarie from it.

For guidance in governing for a short term of four years, Brisbane had his own shrewd common sense and Mr Bigge's comprehensive reports. Amid and despite the machinations of a rising bureaucracy conscious of the pleasure of power, Brisbane found time for astronomy as well as for administration, and he boldly constructed an observatory beside the vice-regal residence at Parramatta. Nor were the earthly sciences overlooked, for the new Governor established an agricultural college, became the first patron of the NSW Agricultural Society, founded in 1822, and enthusiastically espoused the cause of the few scientifically inclined gentlemen who had just formed the Philosophical Society of Australasia, by accepting their invitation to be president.[4]

Like those before and after him, Brisbane was obliged to be meticulous in the submission of returns, the allocation of manpower and the expenditure of public funds. In March 1825 he sent Bathurst a detailed account of 'the Number of Convict Mechanics and Labourers' in government service. In most departments there were pleasing indications that the convict work force had been substantially reduced since the last days of Macquarie's administration. One of only two exceptions was the 'Botanical Garden' at Sydney, where convict workers had increased from eleven in December 1821 to fourteen in March 1825. The Governor hastened to explain this modest increase before pettifogging questions were asked:

To account for the increase in the Botanical Garden, I have only to state that I have added five acres to the Old Garden; that nearly 3,000 varieties of Grapes, Trees, Fruits and other valuable productions of the Vegetable Kingdom have been introduced and cultivated with success in that Establishment in the above period, which I consider of inestimable value to this Colony ...[5]

The Secretary of State was further advised that such success was due to the labours of 'Mr Fraser, Botanist, highly qualified to do every justice to his appointment from zeal, talent and enthusiasm which has brought him in correspondence with all quarters of the Globe, from which he is constantly deriving benefit to the public Service by importations of new, valuable and varied productions'.[6] Truly it was later observed that the botanist had 'a warm friend' in the new Governor.[7]

Brisbane was obliged to go through the recommendations of the Bigge Reports and indicate what steps he had taken to implement them. One specifically related to the Botanic Gardens. The Governor commented: 'Due attention has been paid to this; the original Garden has been nearly doubled,' and, he reiterated, '3,000 varieties of exotic plants, Grasses, bulbs, fruits and vegetables have been introduced within the last year'.[8] Brisbane further encouraged the development of the Gardens by one of his last official acts in appointing Thomas Graham as Assistant Superintendent as from 19 November 1825. At the end of March 1829, however, Graham resigned to manage his own garden and nursery at Chowder Bay, but the position was established, and John McLean immediately replaced him.[9]

To the delight of some of his critics, Brisbane preferred to live in Parramatta, close to his beloved observatory, rather than in the decrepit Government House in Sydney, close to his peevish bureaucrats. He would not have missed a mere 5 acres of the 'Governor's Demesne' as would have the possessive Macquarie, but in any case, both Gardens and Domain now comprised discernible parts of the same Government reserve. It seems, however, that during the administration of both Governor Brisbane (December 1821 to December 1825) and Governor Ralph Darling (December 1825 to October 1831) the Domain, as distinct from the Gardens, received attention rather different from that given in Macquarie's time.

In July 1823, and subsequently, sandstone 'of an excellent colour' was quarried in the Domain, with some offered for sale, and some intended for a new Government House.[10] In January 1825 'some base incendiary' fired 'the brushwood on Anson's Point' thereby threatening 'the whole of the shrubberies on that beautiful scite [*sic*]'[11] and towards the end of the year, Brisbane 'most obligingly lent the use of the Government Domain and stables ...' for the temporary accommodation of the newly-formed Australian Agricultural Company's sheep, horses and cattle,[12] an indulgence which would have delighted Mr Bigge as much as it would have appalled Macquarie. During 1827 and 1828, Edward Smith Hall, outspoken editor of the *Sydney Monitor*, criticized the deterioriation of the walks and fences in and around the Domain, thereby motivating the *Sydney Gazette* to refute such 'doleful laments' and insinuations of neglect.[13]

There were other problems, too, for in August 1828 a party of Aborigines, accompanied by 'a congregation ... of vagabond dogs' set up camp in the Domain, and a few months later their camp fires spread 'with an alarming rapidity and had they not been extinguished by a vigorous effort on the part of Mr Fraser, the whole Domain would soon have been enveloped in flame'.[14] However, sighed the *Gazette*, no action should be taken to expel the incendiaries as 'after all, the poor blacks are the legitimate lords of the soil'.

A far less charitable attitude was adopted towards the packs of marauding dogs, and Mr Fraser advised in the *Gazette* during January 1828 that 'all Dogs found in the inner Part of the Domain around Government House' would be destroyed.[15] There was further concern in May 1828 over the old problem of Domain-dwelling desperadoes: 'The Government domain is ascertained to be a haunt in which runaway prisoners of the Crown evade the vigilance of the Police,' declared the *Gazette*. In the previous month, fear was so aroused that a man walking near the Domain quarry felt menaced by a ruffian armed with a stout stick and a large horse-pistol. Mr Fraser was sought to help give chase, but the ruffian turned out to be a constable, and not 'the murderer Regan' as suspected.[16] Late in 1831, the Domain was still considered to be unsafe for young ladies owing to 'lurking scoundrels'[17] and during the following year official concern was revived over the rather ludicrous situation wherein convicts escaped from government custody to find refuge in a government reserve.[18]

Yet another difficulty arose in 1829 when the area under Fraser's jurisdiction was threatened by a proposal to lease or sell waterfront blocks more especially on the western side of Farm Cove for the wharves, warehouses and dockyards of Sydney merchants, but fortunately the Surveyor-General, Major Thomas Mitchell, had second thoughts and decided after 'further examination of the lower part of the Government domain' that 'there appears to be no necessity for encroaching much upon it for this purpose'.[19] However, the Surveyor-General did select a site for a new Government House just to the north of Macquarie's stables, (the present Conservatorium of Music), and 'the Sale of the Water Side allotments in the Government Gardens' would help finance the project. In November 1832 Governor Bourke agreed to 'the Surrender of the ground required for the New Quay and Buildings' on the eastern side of Sydney Cove.[20] Thus another step was taken in the process of alienating the original vast Crown reserve delineated by Governor Phillip.

Meanwhile, in London, it was considered both scientifically and economically desirable to know just what was being accomplished in that inner sanctum of the Government Domain, the Botanic Gardens. Five days before Governor Darling assumed office in Sydney on 19 December 1825, Earl Bathurst penned a request for regular half-yearly reports on the Gardens, 'for the information of His Majesty's Govern't'. The first report was to include 'an accurate description of the Plants and Vegetables, which are peculiar to the Climate of New South Wales, as well as those, peculiar to other Countries, which are susceptible of cultivation to any useful purpose, if introduced into the Colony'.[21]

On 11 January 1828 Darling forwarded a report,[22] doubtless based upon a 'List of Fruits' and a 'List of Exotic Forest Trees cultivated in the Botanic Garden' compiled in November and December 1827. These lists are interesting as they indicate not only vernacular and botanical names, but also the 'state of acclimatization' and 'By whom Introduced'—for example, Colonel William Paterson, Commissioner J.T. Bigge, Sir Thomas Brisbane, Alexander McLeay, Archdeacon T.H. Scott, Chief Justice Forbes, Edward Macarthur and other notables, as well as overseas correspondents.[23]

Four years later, Viscount Goderich, then Secretary of State for the Colonies, testily advised Governor Bourke that this Report of 1828 was the last received, and he hoped 'to be relieved from the necessity of repeating the Instructions which have been already sent out to the Colony upon this subject' by his predecessor, Earl Bathurst.[24] This was all rather unfortunate for clearly a Report for 1828 was drafted,[25] and presumably sent in 1829; another was compiled by Fraser for the first half of 1830,[26] and yet another, apparently for the second half of that year was 'in the absence of Chas Fraser' signed by John McLean,[27] who also compiled a Report for 1831.[28]

Fraser's Report to Governor Darling in July 1830 reflected much activity during the previous eighteen months. Twenty-six 'cases' or 'collections' of seeds had been despatched to Gardens at Glasgow, Edinburgh, Liverpool, Manchester, Marseilles, Batavia, Mauritius, Ceylon, Calcutta and Hamburg, and to the Horticultural Society of London, the Duchess of Athol, Drummond & Co., 'besides many others'. Dried specimens had been sent to Glasgow (1800), Edinburgh (1200), 'Mr Cunningham, 300 specimens, Mr Telfair, Mauritius, 300 specimens, Botanic Gardens, Liverpool, 300 specimens, besides many others ...'. Cases of living plants had been sent to Gardens at Glasgow, Edinburgh, Liverpool and Batavia, to Charles Telfair at Mauritius, and there was '1 case now ready for shipment for His Grace the Duke of Wellington'. Despite 'the absence of a scientific clerk', Fraser submitted a commendably full account.[29]

John McLean's other Report for 1830 noted that the Garden had

... received a vast accumulation of Plants from various parts of the Globe, more particularly from China, Ceylon, Calcutta, Madagascar, Mauritius, Cape of Good

Hope, Marseilles, Jardin des Plantes Paris, the Royal Gardens of Edinbro' and Glasgow, the Botanic Garden at Liverpool, and many other Botanical and Horticultural Institutions, for all of which reciprocal returns have been made from this Establishment.[30]

Certainly botanists and horticulturists throughout the world, if not the gentlemen of the Colonial Office, were being made aware of the active existence of the rapidly growing Botanic Garden at Sydney.

It is interesting that the Report for 1828 should have referred to the first Report of the previous year, and noted the disastrous effects 'amongst the Herbaceous Plants & Grasses' of 'the unusual drought which has been experienced for the last three Seasons'. These earlier Reports of the 1820s and 1830s principally comprised lists of plants introduced to the Gardens from elsewhere in NSW, the other Australian colonies and from overseas, with indications of how the exotics were faring in their new environment. Special mention was made of those exotics which, like the 'additional Varieties' of olive 'obtained from Europe' by Mrs Darling, had some commercial promise.

Whether or not these Reports were duly received at the Colonial Office, the fulfilment of the bare official requirements would have conveyed little information about the efforts and achievements of Charles Fraser and his labourers, both willing and unwilling, in the unpromising area around Farm Cove, and no information at all about the reputation that Fraser and his Gardens enjoyed within the colony itself. The *Sydney Gazette*, still a virtual organ of the Government, noted in May 1825 that 'we have the authority to announce, that the Botanical Garden has been established with the liberal view of benefiting, not the Government, but the individuals under so considerate an Administration ...' and there was a warm reference to 'our indefatigable Colonial Botanist'. The *Gazette* took a similar line when the *Monitor* complained that the Gardens and Domain were 'occasionally locked' except to privileged military officers. It was explained that on one Sunday the gates had to be locked because the 'immense numbers of individuals' overwhelmed the obliging Mr Fraser, who after all, was entitled to enjoy a Sabbath rest like anyone else.[31]

In October 1827 the *Australian* presented a glowing, if verbose, report which would have given the Whitehall officials more insight into Sydney's botanical establishment than any formal report, as well as an insight into colonial journalism:

A considerable degree of improvement has been of late visible in the Government Botanical Garden, of which Mr Fraser is botanist—an appearance of improvement not alone, owing to proper care being bestowed on the gravelling and cleanliness and just allignment [*sic*] of the walks, but in a great measure owing to the judicious way in which have been disposed the better portion of that variety of rare and beautiful exotic, as also indigenous plants which are abundantly distributed throughout the garden—whose glowing colors and luxurious beauty, the genial and accommodating disposition of our Sydney climate, is helping to expand, as if already in the full pride and possession of their native tropic.

The season is now fast approaching, when madam nature begins to resume her freshest, gaudiest robing ... when a stroll through the ... botanical garden, will not be more grateful to the insipid lounger, who feels his sole wish temporarily gratified in plunging from the broad glare of a fervid sun to cool walks (grateful vicissitude!) and arching trellis-work,—than to the studiously disposed scientific visitor, who loves to explore, admire, and develope [*sic*] nature and nature's work ...

The ebullient tribute also drew attention to some aspects of the design of the Gardens. Furthermore, the visitor was able to observe not only the novel and ornamental, but also the edible and otherwise economically desirable. Among those 'which appear to succeed best' were the banana, mango, mangosteen, plantain, orange, prickly pear, Indian fig, aloe, cocoa, cashew nut, cork tree and India rubber plant, and one could also see tea, coffee and cotton growing. But, alas, there was one grievous shortcoming:

A great portion of the pleasure and improvement to be derived by a visit to those gardens, is in a great measure lost from the want of means of whereby to distinguish one plant from another. Attaching to any remarkable, foreign or native production, a card, having described on it, briefly, its generally received name, nature, quality, &c. would not, we think, be attended with much loss of time, nor with a vast deal of trouble, but be productive of a very general satisfaction.

Nevertheless, 'The attention paid to strangers, and the accommodating disposition of the Colonial Botanist and his subordinates, merit every hearty commendation'.[32] Even Mr Hall of the *Monitor* knew of 'no public officer more entitled to public esteem … His works shew his zeal for the Government he serves, while his urbanity and attention to visitors and persons who desire to make use of his labours and propagate valuable grasses, plants and trees, command universal respect'.[33]

James Atkinson, the successful practical farmer of 'Oldbury' near Berrima, expressed similar views on Fraser's industry and contribution. His 'indefatigable exertions in introducing and propagating new fruits and other productions, and liberality in distributing them among those persons likely to take care of them, are deserving of the highest praise'.[34]

The Gardens' functions as nursery, acclimatization centre and distribution point were long seen as highly important. Settlers wishing to plant ornamental 'exotics', or introduced pasture grasses, fruit trees, grapevines, flax, vegetables and kitchen herbs were periodically advised through the press to call on Mr Fraser.[35] In February 1829 the *Gazette* most strongly urged people to avail themselves of Fraser's supply of 'upwards of 800 layers, and cuttings without number' of some twelve or thirteen varieties of olive which had been 'reared with great success'. Here was a chance for all to help fulfil 'the patriotic wish of His Excellency Governor Darling, to promote … the multiplication of Colonial productions …' and thereby correct the situation wherein 'the balance of trade' was 'fearfully against us'.[36]

There were frequent official demands for Fraser's services. For example, between 1828 and 1831 there came urgent requests from Norfolk Island for cotton seed and for enough seed to produce 15 acres of turnips; from Port Macquarie for seed of clover, grasses and vegetables, especially of garden peas; from Port Stephens for fruit trees and vegetable seeds; from Mr Blackett, Superintendent of Rooty Hill for 'garden Seeds' such as carrot, white and Swedish turnip, celery, cauliflower, 'brocoli', onion, leek, radish, lettuce, 'early York & Sugar Loaf cabbage', kidney bean, white pea and broad bean; from Melville Island and King George's Sound for garden seeds;[37] from Emily Morisset, wife of the Norfolk Island Commandant, for some fruit trees, sage and thyme and a China rose; from the Colonial Secretary for olive layers or vine cuttings to approved settlers such as Sir John Jamison of 'Regentville', Rev. Mr Keane of Bathurst, and Chief Justice Francis Forbes; and from Captain Henry Smyth, Commandant at Port Macquarie, for some more garden seed, and for some mixed hemp and canary seed, as 'my little Favourite Goldfinch is falling off for want of his natural food', with the remorseful plea, 'pray excuse my being so very trouble-some to you ...'. Direct from Government House came a request that four Norfolk Island pines be planted 'within the enclosure of the monument' to the French at La Perouse, and the advice that 'Mrs Darling will feel obliged to Mr Fraser if he will send in her Name some Vegetables and Fruit on Board the "Crocodile" when Captn Mon-tague Sails Saturday Afternoon'.[38] Not only did the governor's lady phrase such cour-teous requests, but also she was 'perfectly accomplished in music',[39] painted wild-flowers with some skill,[40] and, as indicated, imported plants, including olives. On or by 27 October 1830 'Mrs Darling's Border in the inner Domain' was sown with forty-nine species representing remarkable diversity from the 'Large Melon', 'Pum-kin' and coffee plants to *Bauhinia, Crotalaria, Convolvulus,* camphor laurel and mango.[41]

While records of seeds and plants which were received, collected, despatched or sown provided substance for the compilation of the biannual reports, there were other administrative tasks, such as ordering supplies, requesting building maintenance, deploying staff, and seeking to have boards of survey convened to assess the state of repair of items of equipment. As late as mid-1831 Fraser was still being reminded of the procedures to be followed when making estimates and submitting returns.[42]

Attention to these and innumerable similar requests comprised only part of Fras-er's duties. He was also required to request material from overseas, to maintain cor-respondence with eminent botanists throughout the world, and to collect indigenous plants for the adornment of his own gardens and for the augmentation of gardens and herbaria elsewhere. Thus he received 'Boxes of Earth from King George's Sound', had manure carted to enrich the rather sterile area under his care, and checked with some eagerness the consignments unloaded from incoming ships, such as the *Bengal Merchant* which in September 1828 arrived with seeds of culinary herbs and such kitchen crops as radish, lettuce, kohlrabi, leek, melons and tomatoes, and, aston-ishingly, New Zealand Spinach, *Tetragonia tetragonioides,* the edible plant which still grows wild around the shores of Port Jackson. The same vessel brought gardening implements which are worth noting because of the horticultural technology they indi-cate:

12	Spades of Sizes	2	Grafting do.	12	Russia Mats
12	Rakes do.	2	pair Shears.	12	Bell glasses
12	Dutch Hoes do.	1	Do. edging do.	1	Three Gallon Copper
12	Drag do. do.	4	Edging Irons		Still
12	Pruning Knives	6	Hand glass frames	6	Waterᵍ Pots[43]
2	Budding do.	1	Garden Engine		

Fraser was commended 'for the Culture of Colonial Cotton', and for the production 'of excellent opium' which when 'tried by some Gentlemen of the profession' was given 'a superior character'. Another aspect of Fraser's work said to be 'deserving of the attention of our medical gentlemen' was his distillation of eucalyptus oil, probably in the above mentioned 'Copper Still'. It was well that in August 1826, the Agricul-tural and Horticultural Society of NSW should have recommended that the Col-onial Botanist be awarded 'the lesser Gold Medal' for his labours.[44]

In January 1828 Fraser sought an increase in salary as there had been no increase since 1825, when

... my duty was confined to the Garden, the collection at that period did not amount to more than two thirds of its present number, added to which there is now in operation the Colonial Garden, for the reception of Indigenous Plants, the collecting of materials for which will add, much to my Labour and personal expense, Independant of which I have the Superintendance of the Govt. Domain.

Of course, the Colonial Secretary, Alexander McLeay, sought further details, which Fraser immediately provided. He had been appointed Colonial Botanist at 'a Salary of Five Shillings Sterling per diem, and a single Ration' which had been increased to seven shillings a day 'without any allowances but Quarters'.[45] Governor Darling managed to have the salary increased to £150, including 'allowance of sending home seeds, etc., with apartments', while the assistant superintendent was to receive £80.[46] In 1829 Darling attempted to have Fraser's salary increased to £200, but this was firmly disallowed in a peevish despatch.[47]

Fraser's multifarious duties naturally required a workforce which in turn added the chores of gaoler and convict overseer to his other tasks. In July 1828 Alexander McLeay advised

that the Establishment of Convicts to be employed in the Botanic Garden, and the Grounds usually denominated the Government Domain, is fixed as follows, viz.

Watchmen	-	3
Gate Keeper	-	1
Carters	-	2
Overseers	-	1
Labourers	-	10
		17

In addition, 'Nine Invalids are employed, who answer the purpose of more useful men, in sweeping the walks, and other light work, about the Grounds'. To increase this 'fixed Establishment', the Colonial Botanist would require 'special orders'.[48] He was allowed two more gatekeepers[49] and the use of a 'Bullock & Water Cart', although the need for adequate watering facilities long remained. During 1829 Fraser was assigned four to six 'Orphan Boys' who had to be issued with frocks and trousers of 'Russian Duck or Canvas' in the summer, and a woollen supply in winter, with appropriate shirts, shoes and caps, plus an allowance of whale oil for light.[50]

On 24 June 1829 Fraser requested that the Botanic Garden be further enlarged, and within six weeks he was 'to point out the area to be enclosed'. The consequent clearing, quarrying, blasting and carting placed an additional strain on the convict workers and Fraser hoped that 'His Excellencey [sic] will see their inadequacy to perform the necessary Labour required the Garden alone occupying a space of 24 Acres or thereabouts.'[51]

Throughout Darling's time, Fraser proceeded to develop the area adjacent to the shoreline as the 'Lower Garden' which the enlightened Elizabeth Macarthur described as being 'laid out after the plan of the "Glasgow Botanical Garden" of Dr

Hooker—and will be very beautiful—the introductions from Moreton Bay promise to be very ornamental—it assumes already a very tropical character.'[52] This new garden thus demonstrated the advantages of Fraser's frequent correspondence with overseas botanists, notably William Jackson Hooker, then Professor of Botany at Glasgow, and also of his frequent excursions.

Hooker was pleased to publish Fraser's 'Remarks on the Botany, &c. of the Banks of the Swan River' and other western localities, and his 'Journal of a Two Months' Residence on the Banks of the Rivers Brisbane and Logan' in the first volume of *Botanical Miscellany*, 1830. Fraser collected around Moreton Bay and laid out 'the New Garden' at Brisbane, and relied on the commandant for further material for Sydney. In May 1831 Fraser sent 'a collection of vegetable seeds' and some sweet potatoes for 'the Government Garden at Brisbane Town' and requested 'in return ... a few boxes of ... indigenous shrubs or trees'. The suggested method of transport is interesting: 'they may be packed in common cases in dry earth, horizontally, placing alternately a layer of dry earth and a layer of plants, each plant being taken up with a ball of earth, the whole to be covered up as a common packing case, and stowed in the hold'.[53]

In 1829 Fraser's 'Catalogue of Fruits cultivated in the Government Botanic Garden at Sydney' advised readers of J.C. Loudon's *Gardener's Magazine* that custard apples, pineapples, apples, pears, peaches, lemons and grapes were growing, if not thriving, at Port Jackson, and he also prepared 'Memoranda of Australian Fruits and Vegetables' for Ralph Mansfield's 1831 *Australian Almanack* for the guidance of colonial horticulturists. About this time, however, John Henderson from the Bengal Medical Establishment revealed he was hostilely disposed towards several aspects of the role of colonial botanic garden as he saw them:

By all means, let the Governor have his fruits from his orchards; let the burgesses have their promenades in gardens, laid out at their expense. These may supply plants for the greenhouse, or dried specimens for the cabinets of the curious, converting the useless produce of the soil into sources of emolument; but the above are not the object of a Botanic Garden, which is a public institution, created for the purpose of introducing and improving every species of cultivation, which may tend to increase the wealth and prosperity of the state ...[54]

Some of Mr Henderson's critical and pragmatic views were soon to be revived amid considerably controversy, but amicable Charles Fraser would be spared this painful experience.

On 15 August 1831 came the direction that 'you will cause what work remains to be performed on the Walks and Roads in the Domain to be completed by Saturday next 20th instant, in order that the grounds may be immediately afterwards thrown open.' Five days later, Fraser advised that 'Roads in the Government Domain are now made passable for two Carriages',[55] and on 9 September the Governor proclaimed 'that the Grounds in the Government Domain, near Anson's Point, have been laid out in Walks for the recreation of the Public; and that the Domain will be opened for Carriages on Tuesday next, the 13th instant'.[56]

Fortunately there were compensations for Mr Fraser, for on 4 October 1831 Alexander McLeay sent the welcome advice that in consequence of the addition made to the Botanical Gardens, and the laying out of the New Walks in the Domain, for the Recreation of the Public, the Establishment to be allowed for your Department, is to be as follows, viz'—

1	Constable
1	Overseer
15	Gardeners & Laborers
12	Invalids for the purpose of sweeping the Walks and other light work
2	Carters
3	Watchmen
3	Gate-keepers

Total   37 [57]

Presumably the orphan boys remained as well.

Fraser did not have long to regret the disallowance of a higher salary or to rejoice in the allocation of an increased labour force, for when returning from Bathurst with yet another shipment of living plants in December 1831, he became ill at Emu Plains. Further reduced by the summer heat, he just managed to reach Parramatta, where he died on the 22nd, aged forty-three.

On Christmas Day the Reverend Samuel Marsden conducted the funeral service in St John's cemetery, Parramatta, where a broken monument still reflects the doubt evident during the botanist's lifetime about the spelling of his name. Agreeable, industrious Charles Fraser, the grey-eyed 'botanical soldier', subscriber to the Princess Charlotte Memorial, the Scots Church, the Benevolent Society and the Poor Debtors of Sydney Town, died intestate and in debt, but not unsung.[58]

Mr Hall's *Monitor* published a tribute from some friends, perhaps those who had rallied to pay for the funeral. Charles Fraser, they asserted, had risen

... rapidly through his scientific acquirements, self-attained, the urbanity of his manner, and his universal and unremitting benevolence not only to the respectable post which he held ... but to a possession of the full esteem and regard of every individual of his extensive connection of friends and acquaintances. He advanced the cause of science by numerous discoveries in his profession, and his extensive scientific correspondence, has left behind him a lasting monument of his talents and taste in the beautiful gardens, for which, from their extensive utility, the whole Australian public are indebted, but the Public of Sydney in a particular manner; not to mention the roads and walks of the Domain, the last work of his planning, and which might vie in every aspect with almost any other work of a similar description.[59]

The *Australian* was similarly commendatory: 'There have been few men less beholden to schooling, who have displayed so cultivated an understanding, and so enlarged a knowledge of the branch of natural philosophy to which his pursuits in this Country were principally devoted.'[60] Probably the tribute which Fraser would have most appreciated was paid more than thirty years later, when he was mentioned more than 220 times in George Bentham's *Flora Australiensis*.

Fraser's deputy, John McLean, appointed Acting Superintendent from 1 January 1832, bought Fraser's books for £15,[61] and supplied the Report on the Gardens for the latter half of the year. This listed the 'supply of Fruit Trees and Ornamental Shrubs to various parts of the Colony', including apple, pear, peach, nectarine, plum, fig, cherry, lemon, and orange. It was also noted that 'the Plantation of Bananas in the Lower Garden ... produced some excellent fruit'.[62]

It was believed that Thomas Graham, the former deputy, might return to take office,[63] but he apparently preferred to remain at his own establishment across the harbour. Governor Richard Bourke, who had arrived less than three weeks before Fraser's death, was at a loss 'to find any Person in the Colony of Sufficient Science to Succeed him', and advice was sought from the celebrated Robert Brown, botanist on Lieutenant Matthew Flinders's *Investigator*, some thirty years before.

Bourke himself appreciated that 'the utility and success of the Establishment must entirely depend upon the Competency of the person appointed' and such a person would hardly be attracted to the post by anything less than £200.[64] In June 1832 Alexander McLeay made his own plea to the Colonial Office on behalf of the local scientific cause he had long supported:

I hope you will soon send us a *Botanist* as well as the *Zoologist* whom we have been so long expecting ... If no Superintendent of our Garden should be appointed before you receive this I beg leave to submit to you whether Mr Cunningham who passed several years here as Botanical Collector might not be invited to accept of the situation.[65]

Actually, Allan Cunningham had returned to England in July 1831, and in January 1832 was elected a Fellow of the Linnean Society of London. By the time McLeay made his suggestion, it seems that Cunningham had already been approached, but had declined the Sydney position in favour of his brother Richard. On 7 May 1832 Allan wrote a warm commendation of his brother, and asked that he be recommended to the Secretary of State for the Colonies, Viscount Goderich. Allan advised that Robert Brown of the British Museum and William T. Aiton of Kew would be referees for Richard, who had 'been in the service of His Majesty ... for upwards of twenty years' in both 'office and Library in the Royal Botanic Garden at Kew'.[66] The very next day, this letter was favourably endorsed, with the proviso that Brown should approve. This he did, a couple of days later: 'I am able to recommend him as perfectly competent ... both from his knowledge of practical Botany generally, and from his extensive acquaintance with the Plants of New South Wales in particular ... I think his appointment will prove highly advantageous to the establishment'.[67] Goderich promptly appointed Richard Cunningham as Colonial Botanist and Superintendent of the Sydney Botanic Gardens at £200 a year.[68] Alexander McLeay's old friend, Dr W.J. Hooker confided that he did 'rejoice most heartily' at this appointment,[69] no doubt recalling with pleasure those many parcels from Fraser which had contained 'always much that is new, and much to study'.[70]

As a dutiful elder brother, Allan assumed oversight of the whole business. He maintained a vigorous correspondence with the Colonial Office on Richard's behalf, indicating that the new Superintendent wished to take plants with him for the Sydney Garden in boxes which were to be stowed 'as far aft on the quarter-deck or Poop as may be deemed convenient'.[71] Leaving nothing to official discretion, Allan even stipulated that four 'open cases of plants' were to measure three feet by two, and three feet deep, but such was the response from various institutions and nurserymen, that eight such cases were required.[72] On request, Allan also supplied 'a Memorandum relative to the Colonial garden at Sydney ... from which perhaps such instructions, as it may be considered necessary my brother should be furnished with, for his guidance, might be framed'.[73]

Allan Cunningham's experience in the colony both prompted and enabled him to make the most of this opportunity to proclaim his concept of a properly constituted scientific botanic garden. It should be so managed that the Superintendent had ample scope to make a scientific contribution to botany, rather than a gastronomic contribution to officialdom. Charles Fraser had moved in the right and scientific direction by establishing correspondence with 'very respectable institutions' overseas, and by raising

... the seeds of many new and interesting plants, which he had met with in the distant Interior ... a portion of the ground was cleared, for the reception of the indigenous novelties of the inland Country, and thus was the first step taken to convert a kitchen garden into a public botanic-ground.[74]

It was 'presumed' that the Garden had to be '*valuable* to the Colony' but this would eventuate only if the Superintendent's duties included maintenance of a wide correspondence with overseas botanical institutions and studies of plant propagation and acclimatization. The various legitimate functions of the Garden might be fulfilled if there were 'a proper subdivision of the grounds ... into what may be properly denominated, botanic-grounds, and experimental or nursery-quarters—a portion may also still be retained for ... esculent or culinary vegetables ...' and 'above all' space was required 'for the propagation of the many valuable varieties of the *Vine*' then being assembled from Continental vineyards for transmission to Sydney 'by a Gentleman' whom we know was James Busby.

The Superintendent should also undertake the botanical classification of the plants in the Garden, either 'by grouping the species agreeably to what Botanists term the 'Natural arrangement' ... or they may be disposed in ... the Botanic division of the Garden, according to the Linnean method ...'[75] Furthermore, he should 'make periodical excursions to the inland districts, as well as to the out settlements' in order 'to advance our knowledge of the Botany of New South Wales, and in order that the Garden may be occasionally enriched with new species of the indigenous vegetation'. Discovery of sources of 'scarcely known' timbers was especially desirable.

When all this had been accomplished, a 'Catalogue of the Collection' should be compiled so that colonists and overseas correspondents would know what useful plants were available for distribution or exchange. Further, a '*Herbarium* of the plants cultivated in the Garden' should be formed so that 'reference might at all times be made' to it. The implementation of such suggestions, claimed Cunningham, would enable the Sydney Garden to 'rank with those of Calcutta and Mauritius'.

As a final plea for his brother, Allan also suggested that the former Superintendent's house, 'which was originally, most probably simply intended as a vegetable or seed room, and not to be a dwelling-house' was small and very inconvenient, and in need of extensive renovation and enlargement.[76]

Allan Cunningham's memorandum, the most significant document in the early history of the Sydney Gardens as a scientific institution, was adopted—with the nice bureaucratic amendment that the suggested submission of 'an occasional Report' was changed to 'annual'. Once Viscount Goderich relayed the memorandum to Governor

Richard Cunningham who succeeded Charles Fraser as Colonial Botanist and Superintendent of the Botanic Gardens, January 1833–April 1835.
(Mitchell Library, Sydney)

Bourke on 12 August 1832, it had the force of an instruction, and so Allan Cunningham, once dismissed from the viceregal presence 'with a suitable reproof' was now helping to determine official policy.[77]

Allan gave his brother a letter of introduction to his old Parramatta friend, the Reverend Samuel Marsden, declaring that Richard left 'with the best wishes of scientific men in and around London'.[78] Sir William Hooker provided another letter of introduction to Alexander McLeay.[79]

Richard Cunningham arrived in the convict ship *Mary* in January 1833. Within the colony there were high expectations, especially among the growing scientific fraternity, and it was 'hoped from his known talent and assiduity that the colony will soon have a '*Botanic Garden*', in lieu of a repository for turnips and carrots'.[80]

From Fraser's death to Cunningham's arrival, the Gardens and Domain were maintained by the Assistant Superintendent, John McLean and his overseer, Charles Job. By May 1832 McLean had managed to increase the work force to fifty-five, with ten convict labourers and eight temporary hands above the establishment of thirty-seven. Naturally this laudable achievement was questioned, and the additional eighteen were ordered back to the barracks.[81]

In July 1832 Governor Bourke 'paid a visit to the Botanic Garden' where 'the extensive improvements made ... within the last two months, seem to meet with His

Excellency's approval'. Actually His Excellency had not been greatly impressed by the reserves adjacent to the viceregal residence. He confided to Viscount Goderich:

Your Lordship should be informed that exclusive of a small Kitchen Garden the Land about 47 acres to be enclosed as the Government Grounds, contains nothing that can be turned to any profitable use, being almost wholly rock and scrubby underwood. It scarcely affords the maintenance of three Cows.[82]

McLean continued the practice of supplying plants, but sought guidance to meet 'the numerous applications'. The guidance was clear and simple—all applications for seeds and plants had to be made in writing, and all distributions had to be registered in a book.[83] Several of these registers still exist.

Within the Gardens, McLean had the advantage of 'two Cast iron Cylinders ... converted into Rollers'; within the Domain 'on the Government side of Wool-loomooloo Bay' there was to be a new attraction, for Mrs Bigge had received the Governor's approval 'to establish a public Bathing Machine'.[84]

Richard Cunningham took lodgings in Sydney while the 'Garden House' was reno-vated, and appraised the state of the Gardens he was to administer. Not all of his immediate duties seemed directly related to things botanical. At the end of January 1833 there was the matter of slop clothing for the Orphan School apprentices. The order was approved by the Governor, but with the suggestion: 'would it not be better to get rid of the Boys?' At the end of October they were withdrawn, for it seemed 'objectionable to make these apprentices work with Convicts' and in any case, they had 'not generally turned out well'.[85]

Cunningham quickly demonstrated that 'assiduity' with which he was credited. He went to Illawarra for seeds and plants (taking care, doubtless on Allan's advice, to requisition an 'able-bodied Horse & a light Cart from the Carter's Barracks'); ac-quired from 'among the effects' of Dr John Lhotsky, 'an excellent Botanical Press with patent screws, 22 inches long' deemed to be 'a most useful appendage ... in preparing for the Herbarium'; arranged to purchase a covered cart and a bay mare at a sale at Grose Farm; ordered the construction of plant boxes; made an excursion to Emu Plains for seeds and plants, and arranged to visit New Zealand in the *Buffalo*, all within his first year.[86] Material was duly sent to the Royal Gardens at Kew, the Physic Garden at Chelsea, Botanic Gardens at Edinburgh, Glasgow and Liverpool, to the Duke of Northumberland and to nurserymen.[87]

More routine matters included consideration of Mr McLean's application for re-muneration while he acted as Superintendent (a claim which Alexander McLeay felt obliged to disallow although it was 'fully deserved'); 'Hand Carts and Wheelbar-rows', 'Garden chairs and other things' all needed repair; Overseer Charles Job fell from grace 'for harbouring an absentee and reporting him absent on duty' and it was decided 'to appoint in his place ... James Kidd, a steady man employed in the Botanic Garden, at the usual rate of eight pence per diem'; Mr Busby's precious grape vines soon attracted the notice of both humans and birds, requiring the atten-tion of a watchman day and night and a supply of powder and shot; 'the Footbridges and Drains in the Botanic Garden' needed repair, as did the fences; some miscreant stole the 'axles of both Carts' and then the Superintendent of the Carters' Barracks declined to furnish 'Shoes to the Government Horses attach'd to this Department'.[88]

To meet immediate needs and administrative requirements, there were the endless reports, submissions, estimates and requisitions, of which two revealed much about the running of garden and office:

April 1833—from *Wilson and Uther*

4	Garden Rakes	3/4
1	Dutch Hoe	1/1
10	Grubbing Hoes	10/10
9	Breaking up Hoes	16/6
1	p^r Clipping Shears	3/9
3	p^r Edging Shears	7/6
3	Garden Engines	8/12/6
4	Watering Pots	9/0
3	Felling Axes	4/0
8	Pick Axes	14/0
2	Hand Carts	3/15/0
1	Wheelbarrow	4/6

16/3/0 [*sic*]

May 1834—from *Government Stores*

2	reams of brown paper	
2	ditto cartridge paper	
$\frac{1}{2}$	ditto Foolscap do.	
$\frac{1}{4}$	ditto Post paper	
1	lb of Sealing Wax	

Quarter hundred Quills

6	pieces of Red Tape
6	Black & six Red Ink powders
2	quires blotting paper
2	dozen Black lead Pencils
2	Rulers[89]

—from *Mr Samuel Onions*
17 Garden Chairs 8/10/0

Another stationery order included an interesting reference to 'making labels'.

Cunningham's most important task as far as the development of the Gardens was concerned was to prepare a report on their current condition and to make suggestions for their improvement. Cunningham recognized four distinct gardens in his report and accompanying sketch map:

1 'The Lower or Botanic Garden' at the head of Farm Cove;
2 'The old Garden or Fruit Garden' on the east side of the creek on the site of Phillip's farm;
3 'The Experimental Garden' on the western side of the creek in what became (with the Fruit Garden) known as the Middle Garden;
4 'The Kitchen Garden' in the area later known as the Upper Garden.

He recommended the appointment of a regular night watchman for the Lower Garden because of 'recent depredations' and suggested that a sundial 'might with good effect be erected on a central part of this Garden'. The 'old' or 'Fruit Garden' should 'be gradually transformed into a Fruit Garden for the Cultivation of one or more Specimens of each Variety of Fruit suited to the Climate of New South Wales,' and here, too, would be a suitable place for 'some of Mr Busby's French & Spanish vines' to be cultivated and used for the 'grafting of existing unproductive vines'. The Experimental Garden should 'be continued as an Experimental Botanic Garden for the Establishment of indigenous plants of the distant parts of the Colony previous to being removed to their respective places in the Lower Garden'. As for the Kitchen Garden, it was hoped that the Colonial Architect would 'report on the state of the Foot Bridge across the Brook' and upon 'the other Bridges ... with a view to substituting others more safe and ornamental', and that he would prepare estimates for the construction of a stable, a storeroom and a cart shed.[90] This report, the first compiled by a botanist with some professional training, was sent to the Colonial Secretary, Alexander McLeay, in mid-July 1833.

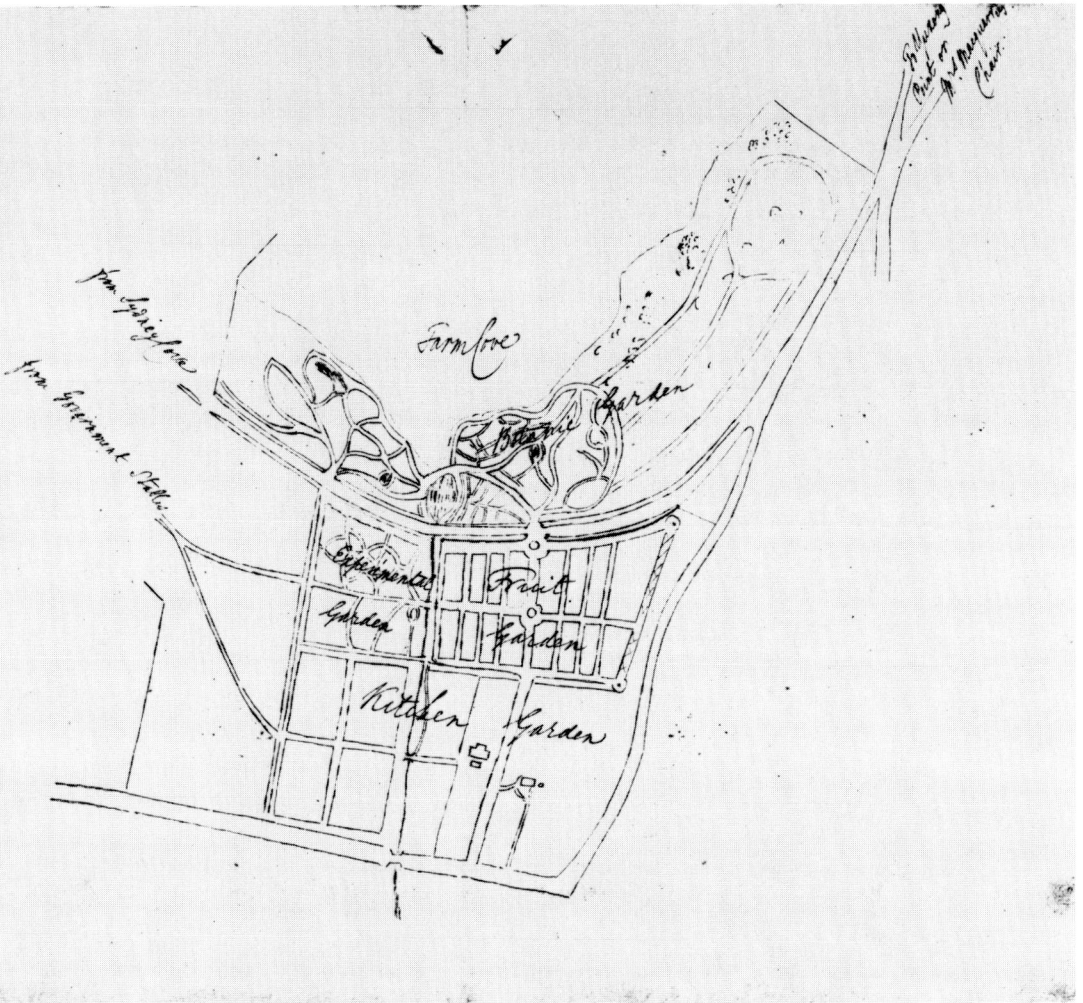

Sketch map which Richard Cunningham submitted with his report and recommendations in July 1833.
It shows the four distinct gardens of the time: Kitchen Garden, Experimental Garden, Fruit Garden
and Botanic Garden. (NSW Archives Office)

Cunningham also sent Mr McLeay a report from Francis Rossi, who with the chief constable, inspected 'the Fences & Railings enclosing the Gov^t inner and outward Domain & of the Botanic Garden' finding some in a 'state of rotteness and likely to fall to pieces by the least touch of the hand'.[91] Other problems were of a more personal nature, for Cunningham himself was taken to task over accounts for accommodation until the Garden House was ready for occupation in August 1833; James Kidd, then a widower with children, sought permission to live in the Domain Lodge; there were constant reminders to submit half-yearly reports; there were problems over 'the admission of persons to the Government Domain and the Botanic Garden' and there were questions about the gatekeeper's entitlement to collect firewood and graze his cows in the Domain and others having to endure his insolence.[92] The collecting trips to New Zealand, Norfolk Island and Van Diemen's Land must have afforded welcome relief.

However tardily supplied, the four reports compiled by Cunningham and McLean for 1833 and 1834 recorded the introduction of paeonies, fuchsias, gardenias, other ornamental plants and about fifty plants from New Zealand, developments in the Lower Garden, and a vigorous programme of distribution 'to the several Government Establishments and private Individuals' comprising over these two years, 2300 fruit trees (including some olives); 800 olive plants; 3520 ornamental plants and flowering shrubs; 3075 grape vine cuttings plus 'a vast number not Counted'. McLean made the interesting observation that 'the Government Gardens of Sydney' then comprised 'in the Whole about Twenty acres under Cultivation, and four more Brokeup for Botanical purposes'.[93]

In the summer of 1834–35 Cunningham visited Mount Tomah,[94] which had so fascinated his brother eleven years before, and which was destined to assume special significance in the extension of Sydney's Royal Botanic Gardens. In January 1835 Cunningham vainly sought to save the errant Robert Owen, Gardens carpenter, from the treadmill and banishment 'to a distant service', before attention was diverted to an urgent request for seeds for a 'Sand Hill at Newcastle ... nearly ten acres' from which sand was blowing at an alarming rate because the shrubs had been 'cut down very improperly'.[95] In February came the fateful message that 'the Surveyor General wishes you to accompany him on his intended Expedition'.[96] Accordingly, in March the Colonial Botanist joined Major Thomas Mitchell's expedition down the Bogan River, where a month later he was reported missing.

It had been Cunningham's practice to diverge from the expedition's line of route in order to cover more country in his search for specimens. Mitchell, whose camps and progress were organized with military precision, did not applaud the botanist's zeal: 'I had repeatedly cautioned this gentleman, about the danger of losing sight of the party in such a country; yet his carelessness in this respect was quite surprising'.[97] An intensive search was made. Guns were fired, bugles were sounded, messages were left on sticks, searchers called out as they scoured the bush, Aborigines were questioned, but all in vain. The botanist's dead horse was found, with his whip, saddle and other items, and some footprints were followed, but there was no sign of Cunningham himself. Mitchell was obliged to proceed, with his anxiety about the botanist 'embittered with regret at the inauspicious delay of our journey which his disappearance had occasioned'.[98] It was later ascertained by Lieutenant Henry Zouch of the Mounted Police that Cunningham, apparently in a delirium, had perturbed some Aborigines by wandering around their camp at night, and was accordingly clubbed to death on or about 20 April 1835.

Once news of the tragedy reached Sydney, John McLean again found himself Acting Superintendent, and James Anderson, who had arrived in the ship *Brothers* in August 1832 with many years of experience as a botanical collector to his credit, was appointed Assistant Superintendent in September 1835. In England, Allan Cunningham was advised of Richard's death and again offered the position of Colonial Botanist. This time he accepted.

One of the more formidable tasks which fell upon John McLean was the request, transmitted through Alexander McLeay from Lord Glenelg, that 'the Superintendant ... send home, addressed to this Office, a supply of one or two pounds of Seeds of 20 or 30 Species of the more ornamental plants of the Colony,—each Packet to be accompanied by directions for ... proper cultivation'.[99] These would be given to the Horticultural Society. In view of the minute weight of the seeds of many of the 'more ornamental Plants', this request was something of an imposition but it was met, at least in part, by March 1836,[100] and in the following month it was advised that provision would be made 'for a Building for drying and arranging Seeds &c to be erected at the Botanic Garden'. Furthermore, as Richard Cunningham had recommended, forty men were to be specially attached to the Botanic Garden and James Anderson was charged with supervising a large gang to trench and plant Hyde Park.[101]

Anderson shortly urged that convicts once assigned to this department should not be withdrawn 'after they had become servicable [*sic*] to the Place'. The point was taken: 'Let them be regularly assigned and Mr Anderson charged to pay them every attention instructing them in gardening ...' They were to be assured that good conduct would be rewarded, while 'idleness and profligacy' would be punished. It is clear that the work force assigned to the Gardens and Domain rose and fell according to officially recognized need. Despite occasional requests for certain men to be appointed, the Superintendents were generally obliged to receive anyone available and to lose anyone required elsewhere or who committed a misdemeanour to be marched off to an ironed gang or a treadmill preparatory to distant assignment.

Early in 1837 there was concern over 'the plunder ... of articles left by the late Mr Cunningham' in the Gardens house,[102] and there came an instruction that communication with the outer Domain via Fort Macquarie should be made 'in a Direction nearer to the Waters Edge than the present Road'. In accordance with proposals for the new Government House, these improvements were to be undertaken without delay. Interestingly, the man chosen to be overseer of this new work was a ticket-of-leave holder, James Tucker, destined to achieve posthumous fame as the probable author of *Ralph Rashleigh*. At the time, however, any literary promise was obscured by consideration of whether he should be paid at 2/10 or 3/3 a day for directing the labours of his gang of twelve.[103]

Accompanied by two precious 'glass-roofed Cases filled with many ornamental European Plants, and a box containing the roots of the several species of European Paeonies' from Loddiges of Hackney, Allan Cunningham, former King's Botanist, returned in the *Norfolk* on 12 February 1837 as 'Colonial Botanist and Superintendent of the Botanic Garden',[104] ready to defend that professional freedom he had sought for his brother. By or about this time, John McLean had left the Gardens to become Superintendent of Agriculture on Norfolk Island, where on 15 February 1840 he was drowned when a boat overturned in the surf. He was forty-three.[105]

Cunningham saw that some things had hardly changed during his six years' absence from the colony. At the outset he drew attention to 'the state of dilapidation of the Cottage of the Superintendent of this Garden'. He sought, and ultimately gained, permission to rent accommodation while repairs were made.[106]

One of the first official tasks was to 'supply from the Government Garden as much fruit and Vegetables as can be obtained for the use of the Emigrants on board the Lady McNaghten now lying under Quarantine at Spring Cove'.[107]

In March, Anderson reported trouble with the impulsive and irascible principal police magistrate, Colonel Henry Croasdale Wilson, who, preceded by a constable on foot, commanded the watchman to open the gates to the Lower Garden, and drove to the upper or south gate and back again. Predictably, the new Colonial Botanist took a stern view:

> I respectfully take leave to observe, that the walks in this garden were not originally constructed to admit of ingress & egress by Horse or Carriage, nor have they ever been, I believe considered, in the state of repair in which they are maintained, other-wise than for the accommodation of visitors on foot, who may wish to inspect the plants, and enjoy in perambulations through the grounds, that rational Recreation which a Public Garden possessing a considerable variety of indigenous and exotic plants, is capable of affording.[108]

There were also many routine matters. Cunningham asked to receive the *Government Gazette* and advised his 'intention to re-establish the correspondence that at one period existed' with overseas Gardens and 'with the several Missionary Stations in these Southern Seas, with a view to mutual benefit, by an interchange of useful and ornamental plants ...'. A further shipment of twenty-four large packets of seeds was prepared for the Horticultural Society, although it was not 'thought necessary to sup-ply so large a quantity as one or two pounds of each'; vines 'from Mr Busby's collec-tion' were sent to 'the South Australian Settlement'; preparation of Hyde Park for additional planting was completed, and Cunningham suggested that 'white cedars, thorn-acacias, oaks and some elms' be procured from the late Thomas Shepherd's Darling Nursery, where 500 trees 'upwards of eight feet in height', twice the size of those in the Garden nursery might be bought for £5 per hundred; a further 'five bundles of Oak-trees' for Hyde Park were received from the Government Domain at Parramatta.[109]

About four months before Cunningham left England in the previous October, Gov-ernor Bourke had taken a step which was to cause some friction for the next fifteen years. Bourke had established 'A Committee of Superintendence of the Australian Museum and Botanical Garden'.[110] Cunningham, wary of Governors and sensitive to criticism, had now not only to satisfy the Governor, but also to keep favour with a new committee, some of whose members were itching to display their own erudition as they pleaded the cause of colonial science. Certainly Cunningham had friends like Alexander McLeay on the committee, but this did not diminish its obvious power over the two institutions under its superintendence, or its influence upon the Governor—one glance at the names was enough:

Alexander McLeay, Colonial Secretary and patron of science to whom the Colonial Botanist was directly responsible.

George Macleay, son of above, pastoralist, naturalist, and companion of Charles Sturt during explorations of 1829.

Charles Sturt, explorer, then living at Mittagong.

Phillip Parker King, naval explorer whom Cunningham had accompanied on several voyages, and prospective member of the Legislative Council.

Sir John Jamison, 'the Knight of Regentville', near Penrith, prospective member of the Legislative Council.

William Macarthur, son of John, one of the recent successors to the Camden Park estate, accomplished horticulturist, viticulturist and amateur botanist.

Alexander McLeay, Colonial Secretary of NSW 1826–37. A Fellow of both the Linnean and Royal Societies of London, he was a foundation member of the Botanic Gardens Sub-Committee. (Historic Photograph Collection, Macleay Museum, University of Sydney)

Edward Deas Thomson, Governor Bourke's son-in-law, very soon to be Colonial Secretary.

*Robert Andrew Wauch, recently arrived from Scotland, soon to be pioneer settler on the Upper Hastings and founder of Wauchope.

*Dr John Vaughan Thompson, Deputy Inspector-General of Hospitals.

*George Porter, a respected publican and businessman of Brickfield Hill, Sydney.

Dr George Bennett, eminent surgeon and amateur naturalist, author, and Secretary of the Committee until 1 July 1841.

There were two sub-committees, one for the Museum and one for the Gardens. Some members belonged to both. The original Gardens sub-committee comprised those whose names are asterisked.[111]

There is little doubt that the establishment of the committee was due to the scientific proclivity of the Colonial Secretary, Alexander McLeay, a Fellow of both the Linnean and Royal Societies of London. McLeay, who arrived in January 1826, had been an influential intermediary between Governors and Colonial Botanists for a decade and more, and had been prominent in the formation of the Colonial (later Australian) Museum. Widely respected for his knowledge of entomology and horticulture, McLeay had established one fine garden 'full of ornamental trees and shrubs'[112] at Brownlow Hill near Camden, while he developed an even grander one of over 50 acres as the setting for his new home at Elizabeth Bay. The gracious Regency house now overlooks a residential area where fragments of the once-famed garden may still be found.

McLeay did not attend the earliest meetings of the Botanic Gardens Sub-Committee. Indeed, until he began to do so, interest seemed to be fading altogether, despite sanguine efforts to call a meeting every week.[113] The first meetings were concerned with the powers vested in the committee and with labour, finance, and the distribution of grape vines to suitable applicants. McLeay attended his first meeting on 30 June 1836 when it was decided 'that a Plan of the Garden should be made and laid before the Committee'. The Deputy Surveyor-General, S. A. Perry suggested that a dozen tracings for committee members might be made 'by means of Zinco-graphy'.[114]

The all-important matter of vine cuttings apparently occupied the next few meetings, but as from 9 July 1836 the sub-committee was left in no doubt about its role. According to McLeay, who drew up the 'Heads of Instructions' at the Governor's behest, it 'should have Superintendence' of 'the principal Garden, the New Garden in farm Cove, and that nearest the Government Stables'; the Kitchen Garden and inner Domain were 'to be kept distinct' and 'under the Superintendence of the Colonial Botanist'. McLeay continued: 'The Colonial Botanist and Superintendent are to follow all such directions as they may from time to time receive from the Committee in all matters connected with the Botanical Gardens'. But care was taken to add:

No Vegetables or Fruit Trees are to be raised in the Botanical Gardens, excepting such as are of so Valuable or Rare a kind, as it may be proper to place there for better preservation. Such of the fruit trees now there are to be removed as the Committee may think proper.

McLeay also appreciated that the establishment of private nurseries now rendered 'it unnecessary to continue the distribution of Shrubs, Trees, &c.' and so it was proposed that the plants in the Gardens should be confined to those 'of a Rare Kind that cannot be obtained elsewhere in the Colony'. In McLeay's view, the committee should 'furnish an Annual Report every May', indicating progress 'made towards a Scientific arrangement of the Plants', together with a census of plants cultivated and an account of exchanges. The first report should include 'a Succinct historical Sketch of the Establishment' and 'some account of what it contained when taken over by the Committee'.[115] Allan Cunningham thus had come to assume control of an institution which had already been 'taken over'.

In June and July 1837 Cunningham dutifully asked the Colonial Secretary (Edward Deas Thomson and a member of the committee) whether he had to obtain the authority of the committee before obliging applicants for seeds and cuttings. He also asked whether he had 'to communicate with the committee of management for such general instructions as may be considered necessary for ... guidance, as Superintendent of this Garden'.[116] Cunningham was informed somewhat coolly that it was 'understood the committee will give their usual attendance for the purpose mentioned', but no hint was given of the instructions to be expected.[117]

As the tension mounted during the latter half of the year, various matters required attention. In June, there came a request for 'a package of Garden seeds the most useful esculents for the use of the Missionary Establishment for the Civilization of the

James Kidd who worked at the Botanic Gardens from December 1830, when he arrived as a convict, until 1866 when he was superannuated as overseer.
(from a photograph of 1863, Royal Botanic Gardens Library)

Natives at Port Phillip,' and in August the committee supported the request of Lieutenant Symonds of HMS *Rattlesnake* for 'such Seeds and Botanical Specimens as may be practicable'.[118] Cunningham also championed the cause of James Kidd, whom Richard had had appointed overseer in July 1833. Kidd, a gardener from Fifeshire, had arrived in December 1830 convicted of forging notes. Allan now recommended that Kidd be entitled to a ticket-of-leave and paid 3/3 day, but the Governor noted that 'only one overseer at eight pence per day' was provided for in the estimates. For one of Cunningham's interests, abilities and temperament, this was all too irritating. Kidd himself held references from Chief Justice Forbes, Archdeacon Broughton, Dr James Bowman and Joseph Montefiore, a well-known merchant.[119] Nevertheless, Governor Bourke still had reservations, while making some concessions in view of Cunningham's appeal:

... at your desire and that of Mr Anderson His Exy will not object to give Kidd a Ticket for Sydney for the time he is employed in the Botanical Garden (but no longer) with pay at the rate of two shillings per day from the date of his obtaining his Ticket with the House & garden he occupies His Exy considers equivalent to the usual wages of a Ticket of leave overseer not possessing those advantages.[120]

The Governor, also revived the matter of training 'young Convicts ... in the art of gardening' so that 'having obtained Tickets of Leave for country Districts they will readily find good employment from the more opulent settlers'. Barracks then being built within the Gardens would soon be completed to accommodate 'from forty to fifty labourers with one or two Overseers'. Mr Anderson seemed ideally qualified to provide the instruction of these convicts 'as well as ... their superintendence'. His salary would accordingly be raised to £100 for this work, while 'the Scientific details of the garden rest with the Botanist'.[121]

On 15 November 1837, five days after this scheme was announced, Allan Cunningham submitted his resignation. There had been difficulties in establishing and maintaining discipline over the convicts and this problem would be increased when the proposed barracks were completed; clearly skills in practical horticulture and landscape gardening were required more than 'the possession chiefly ... of Botanic Science'. Therefore, Cunningham felt 'in a great measure unfitted for the future management of this Establishment'.[122]

Cunningham's resignation was received in the Colonial Secretary's Office on 16 November, but it was not until 4 December that it was annotated. The Governor had been pleased to accept the resignation 'from the 1st of January next'. Meanwhile, would Mr Cunningham 'have the goodness to prepare a list of the different varieties of plants and shrubs which the Botanical garden now contains'. There also came a note that two prisoners in the Garden had been allowed 'the indulgence ... of working on Saturday afternoons for themselves'. This was to cease, and of course a previous instruction still stood: that the Colonial Botanist was not authorized to allow prisoners incentive payments such as tea, sugar and tobacco considered 'likely to induce the Convicts to more exertion'.[123] To the ailing Cunningham, peaceful Kew must have seemed a world away.

Having accepted Cunningham's resignation, Governor Bourke resumed his packing, for he too, had resigned on a matter of principle, and was to leave for England via South America on 5 December 1837. It was a nice gesture that he should shortly send from Rio de Janeiro 'several orange Trees, Shaddocks [grapefruit], Bananas' and other plants for the Garden.[124]

On New Year's Day 1838 Sydney's Botanic Gardens and Domain were officially bereft of a Superintendent. On 4 January James Anderson applied to the Colonial Secretary for the position, while Cunningham wrote from his cottage in Elizabeth Street, advising the committee that Anderson was fully competent for the task.[125] On 22 January Edward Deas Thomson wrote to 'The Late Colonial Botanist' seeking 'with as little delay as possible, for transmission to the Secretary of State, the usual Half yearly report of the Botanical Garden up to the close of last year'. To meet this request, Cunningham agreed to be reinstated temporarily for January, for which he ultimately received £16-13-4. James Anderson was appointed Superintendent as from 1 February 1838, 'until you are either confirmed, or another person nominated ... by the Secretary of State'; James Kidd was granted his ticket-of-leave on 21 December 1837 and was appointed Assistant Superintendent; Cunningham submitted the requested census of plants in the Gardens, and received the thanks of the Acting Governor, Colonel Kenneth Snodgrass.[126] Thus the botanical department was administratively, if not emotionally, stable by the time Governor Sir George Gipps arrived on 24 February.

To his friends in England, Cunningham announced: 'Tell all that I have discharged the Government cabbage-garden in disgust, and am now to enter with all my might, mental and corporeal, on a more legitimate occupation for a few months'.[127]

For the colonists of New South Wales, the *Sydney Herald*, anxious for a final stab at the Whiggish Bourke, provided a much more interesting statement:

We have had frequently to call the attention of the Colonists to the fact, that a Kitchen Garden, under the pretence of being a Botanic Garden, is supported in Sydney at an expense of from £800 to £1000 a-year! But what care the 'powers that be' for that as long as hungry officials can 'furnish forth' their tables with fruit and vegetables grown at the public cost? We scarcely ever walk through this garden without seeing some servant with a basket carrying off vegetables, or fruit, for Mrs This or Mrs That—the wife of some official ... People frequently ask 'What becomes of the fruit and of the vegetables grown in this *Botanic* Garden?' The question is easily answered. They fall into the insatiate maws of all-devouring officials ... It is, in fact, so barefaced that Mr Cunningham would no longer consent to remain a mere cultivator of official cabbages and turnips; and, accordingly, he has resigned the management of the *Botanic* Gardens in disgust.[128]

Gipps thus arrived to find the press and much public opinion in support of the injured Colonial Botanist, who had valiantly defied the attempts of privileged citizens and officialdom to deprive him of his professional function and dignity. For his part, Cunningham had, by 9 February 1838, completed a voluminous Report on the Gardens for the period 13 February to 31 December 1837. He wrote temperately and objectively of the seasons, a drought, Mr Busby's vines, receipts and distributions of plants and seeds, progress with fruit trees, and other routine matters, though there were rather more pointed references to the division of labour in the Gardens. Cunningham emphasized the great importance of correspondence and exchange of material with overseas institutions, but work in the Garden generally was hindered by 'the very inadequate supply of water'. The Garden had no well of its own, and the stream passing through it frequently failed.

Cunningham concluded with the hope that if his suggestions were adopted and developed, the Gardens 'would be fully entitled to the Support of a liberal Government, and would be well worthy of a place among other *scientific* Institutions, of this most rapidly advancing of Her Majesty's Colonies'.[129]

Gipps was concerned about the situation. As he explained to Lord Glenelg, he felt

... a very lively regret at the prospect of losing the services of a person so distinguished as Mr Cunningham for his knowledge of Botany, I could not suffer him to leave the Colony, without making an effort to induce him to change his purpose. I accordingly entered into a negotiation with him, through the medium of the gentlemen under whose management the garden is placed, but I am sorry to say, without success.[130]

Thus the committee which Cunningham saw as a potential obstacle to his work was required to be an intercessor.

Only McLeay, Jamison and Macarthur attended the meeting at which the Governor's wishes were announced. They 'lost no time in communicating with Mr Cunningham' who on 12 March 1838 declared the terms under which he would be prepared 'to accept (for a period) an appointment of Government Botanist in this Colony':

1 A salary of £450 per annum.

2 Free passages to out-settlements such as New Zealand and Van Diemen's Land.

3   Convicts, clothed and victualled by the Government, to assist on excursions.

4   Two tents to be provided (one 'of either Russia Duck or the lightest Canvas' and one of 'the "Parramatta Factory Woollen Cloth".')

5   'Draught horses for each Journey.'

6   'A Strong Cart for the transport of living plants.'

7   Brown paper and other necessary stationery to be provided.

8   'A carte blanche as to the disposal of ... time and going on journeys'.

Cunningham insisted that any future appointment 'will *never* connect me again, with the Botanic Garden, from all the various duties of which I have permanently retired, with the exception of that of being instrumental in introducing desirable exotic plants, and the rarer indigenous vegetables to the care and culture of Mr Anderson.'[131] Sir John Jamison thought £350 would be a fair salary, but his two colleagues recommended the sum requested; however the committee could not agree to the 'carte blanche' request, not because of any lack of confidence in Cunningham, but because of the precedent it would set. It would be better if Cunningham were to spend 'say one half of each year (the seasons to be chosen by himself) in excursions for the purpose of Science'. It was further recommended that Cunningham's 'connection with the Botanic Garden shall be discontinued except in the matter proposed by himself, unless indeed he be made a Member of the Committee established for its General Management'.[132]

Gipps did not reject these terms outright, but sought more information on the equipment which Cunningham claimed would be required for botanical exploration. In reply, Cunningham asked for 'the usual travelling or Bush ration' for his men 'as supplied to surveying parties' and he described the clothing issue, tents, bedding, horses, etc. in similar detail.[133] This quickly brought negotiations to a conclusion. Gipps considered that the demands meant an immediate outlay 'of something more than £200 ... and that the *fixed* ANNUAL expense would *certainly amount to not less than £850*'. Cunningham protested that these amounts had been inflated, but despite his own moderate estimate, the matter was closed. An astute administrator, Gipps held that in Downing Street, 'the governor who keeps his government out of debt is the best'.[134]

Cunningham was by now suffering from advanced tuberculosis, and perhaps was uncharacteristically irascible. The paternalistic interest of Banks and Bathurst had gone and the influence of the Hookers was of a different kind; Gipps was responsible to a distant Secretary of State and to a local Legislative Council, both more sensitive to economic than to scientific issues, and the colony was in a state of drought.

The new Governor felt bound to trust the committee which his predecessor had established and to rely on its recommendations. He stood between a professional botanist, conscious of his previous exploratory work and of his scientific potential, and the representatives of a growing number of enlightened amateurs conscious of the contribution they could make to the development of colonial science. The impasse was complete.

Cunningham sailed for New Zealand in the French corvette, *L'Héroine* on 15 April 1838, while the Governor took care to issue:

... the most positive order, that no culinary vegetables of any sort shall be raised in the garden,—that the fruit-bearing trees shall gradually be removed ... and that the kitchen garden properly belonging to Government House, ... though adjoining the Botanic Garden ... shall be taken entirely out of the management of the Superintendent.[135]

Allan Cunningham and the *Sydney Herald* had scored one victory at least.

Returning from New Zealand in October, Cunningham, though now very ill, arranged to join Captain John Wickham in the *Beagle* bound for the north-west coast. By June 1839 he was so debilitated that his friends took him from the Elizabeth Street house to the cottage in the Botanic Garden, where on 27 June, sixteen days short of his forty-eight birthday, he died in the company of his faithful friend, James Anderson. Lamenting the loss of his old shipmate, Phillip Parker King, a member of the committee and Cunningham's executor, declared: 'Alas, poor Allan! he was a rare specimen—quite a genus of himself; an enthusiast in Australian geography; devoted to his own science, Botany; a warm friend, and an honest man ...'[136] King and others called a meeting in the Gardens for 30 October 1839 to consider a suitable public memorial, and, in accordance with the will, a plaque was to be fixed to the wall of St Andrew's Scots Church. Originally intended to be 'a handsome sepulchral urn upon the small island in the lower Botanical Garden which is surrounded by willows',[137] the public monument ultimately took the form, in 1844, of an obelisk, within which, through the interest of J. H. Maiden, Cunningham's remains were later enclosed. The ledger stone from the original grave is preserved in the wall of the Herbarium.

The obelisk erected in 1844 to commemorate Allan Cunningham became his tomb in 1901. Photograph attributed to Alexander Brodie, about 1870.
(Historic Photograph Collection, Macleay Museum, University of Sydney)

Because of official attitudes and a succession of short-lived Superintendents, the scientific function of the Gardens, so strongly advocated by Cunningham, was not significantly developed for another decade. Meanwhile, Cunningham's 'valuable and extensive herbarium' reached England safely, and passed 'by the kind bequest of the talented and admirable collector' into the possession of his friend and biographer, Robert Heward. There was no place, nor even the need, for such a collection in Sydney until the scientific aspect of the Gardens had been fully appreciated.

In January 1839 James Anderson furnished a report for the latter half of the previous troubled year, recording the usual exchange activities, the distribution of over 3000 more cuttings of Mr Busby's vines and the deployment of staff. Anderson even revived the matter of incentive payments, seeking indulgences of tea, sugar and tobacco to be granted to deserving convicts, say, the fifteen best behaved in the Gardens. Gipps agreed to 'small gratuities' if Anderson could reduce the number of men or otherwise make savings.[138]

Like his predecessors, Anderson went on summer collecting trips to the Blue Mountains, and also like them, he was faced with problems within the Garden itself. While he was away in the mountains on one occasion, the military band 'formed themselves in the Lower Garden and played for a considerable time, which caused the collection of a concourse of People; who not only trampled the Plants but broke down many of the Palings ...'[139] Colonel French was admonished and the musical invasions ceased.

It seems unlikely that Anderson, who was appointed Superintendent, but not Colonial Botanist, assumed much additional responsibility on Cunningham's death, although, when he also died on 22 April 1842, he was courteously, and warmly, described as 'the Colonial Botanist, under whose direction the very many and great improvements effected in the Gardens were made ... Mr Anderson was indefatigable with the pursuit of his favourite science, and was eminently useful in advancing the interests of horticulturists and floriculturists in this colony'.[140] Friends rallied to erect a monument over his grave in the Devonshire Street cemetery. Like most others, it was removed to Botany in 1901, but was in a sorry condition when finally destroyed in 1975. Although 'a worthy friend, a valuable public officer and a zealous and indefatigable Botanical Collector', Anderson was apparently, as Cunningham recognized, an accomplished collector and horticulturist rather than a scientific botanist. Ludwig Leichhardt, the newly arrived Prussian-born naturalist, certainly believed this, and scornfully wrote: 'They appointed an ordinary gardener, a man without scientific knowledge, under whom the garden became nothing but a kitchen garden for Government House.'[141]

During Anderson's brief term, Governor Gipps made three additional and very significant appointments to the Committee of Management[142]—William Sharp Macleay, eldest of Alexander's sons and a competent naturalist; Dr Charles Nicholson, landowner, physician, Member of the Legislative Council and shortly one of the founders of Sydney University; and the Reverend William Branwhite Clarke, clergyman and geologist, soon to be appointed first Rector of St Thomas's, North Sydney. On 1 July 1841 Clarke succeeded Dr George Bennett as Committee Secretary.[143]

At this unsettled time, Gipps received a rebuke from Lord Stanley because Sir William Hooker of Kew had received no acknowledgement either of 'a Pacquet of

Seeds' or of a letter sent to the Superintendent. Stanley pointed out that it was from Hooker, and not from anyone in the colony, that he had learned of Anderson's death, and he tartly concluded 'you will inform me of the name of the person whom you have appointed to succeed Mr Anderson'.[144] Gipps hastened to explain that 'the Botanic Garden at Sydney is under the management of a Committee of Gentlemen, who are for the most part unconnected with the Government,'[145] and referred Stanley's letter to Edward Deas Thomson who passed it on to the President of the committee, knowing that Alexander McLeay would open it. McLeay readily took up the challenge, and sent a chilly reply to Thomson whom he considered an unjustifiably favoured usurper anyway.[146] He knew nothing of the seeds, but would take up the matter directly with his 'very old and particular friend', Sir William Hooker.[147]

Meanwhile, the matter of Anderson's successor had to be settled. Within five days of Anderson's death, applications were submitted by Francis William Newman and James Curnow. Others were received from Nasmith Robertson, John Edwards, Martin Tobin and James Kidd. Ludwig Leichhardt, who on the advice of 'several well-disposed and influential people' and with his 'mind ... teeming with the scientific possibilities' of the position, also applied despite his belief that the salary had been reduced from 'about ... £250 ... to £120, and they even temporised about filling the position at all'. Leichhardt believed he had Gipps's support, but Alexander McLeay, 'the most influential member of the Botanic Gardens Committee ... carried the day with ease and my application was rejected'.[148]

On 12 May 1842, on the motion of Alexander McLeay and William Macarthur, and with the support of Dr J.V. Thompson and the Reverend W.B. Clarke, the committee decided in favour of Nasmith Robertson, to whom, according to Leichhardt, McLeay had already promised the appointment.[149] Certainly Robertson, a native of Kilmary, Fifeshire, who arrived as a steerage passenger in the *Pyramus* in May 1829, was well-known to William Macarthur, having been 'for many years principal gardener' to the Macarthurs at Camden Park, where he 'had the chief direction of their extensive gardens and vineyards'.[150]

The Governor found some difficulties in the committee's decision, noting on 4 June:

I have been waiting for more than a fortnight to see Mr Robertson & have succeeded only in finding him this morning.

His appointment may now go forward though he is a man considerably advanced in years and not likely long to be fit for any active exertions, the Committee must be informed that I cannot guarantee his continuance in Office longer than he may be able efficiently to perform the duties of it ... He is to be called Superintendent of the Botanic Garden—not Colonial Botanist.[151]

Mr Kidd was also a little piqued, and notwithstanding the well-known educational attainments of the committee members, he bravely committed his case to paper:

Having had the entire Charge of the ... Establishment during the perried from the death of Mr Anderson (and from Delicate State of his health Some time preivius) to the appointment of the present Superentendent Mr Robinson [sic] I have the honor most respectfully to request that you well will [sic] Submit for the Kind Consideration of his Excellency this my application for Some portion of the Superentendents Salary during the peried that the Situation was not filled (about Six weeks) which Indulgence I trust my Conduct and unrimitting exertions for the improvements of the Gardens during the time I have been Overseer will in Some Measure be deemed deserving ...[152]

As the sum of £24-8-10 had been saved, the Governor approved Kidd receiving a quarter of it, no doubt considering that this would be something of a boon to a man on 3 shillings a day.

At the end of 1842 Robertson was granted permission to collect seeds around Bathurst and in the Bargo area, to which he returned the following summer. However, fears concerning his health were all too well founded, and like his predecessor, Nasmith Robertson had little time to prove his worth. In June 1844, at the behest of Helenus Scott of Glendon, Leichhardt called on the Superintendent, but considered 'the poor man seems to lie on his deathbed; he suffers of rheumatism and probably of dropsy'.[153]

On 8 July 1844 Nasmith Robertson, now sixty and the recent proud recipient of 'a diploma as Fellow of the Royal Academy of Botany, at Stockholm' on account of his botanical services to Sweden, died in his cottage in the Gardens. Once again the Gardens and Domain were without a Superintendent, but for the ever-available James Kidd.

This period immediately after Cunningham's death was marked by events and conditions which had wide and long-lasting effects on the colony. Transportation of convicts to New South Wales ceased in 1840 after two years and more of drought had seriously affected the wool-based economy. A shortage of labour developed, markets for wool and meat declined, and by 1842, the colony was in the throes of depression.

Like other institutions, the Gardens were affected. At the end of January 1844 Robertson had advised that he could not manage the Gardens at the same level on the proposed budget, which had been reduced from about £800 in 1843 to £634 in 1844.[154] He felt that the number of men would have to be proportionately reduced from thirty to twenty-four as one result of such measures.[155] James Kidd applied for the position in July 1844 during hard times.

There were other contenders, as before. These included Alexander Burnett, W. Armstrong, Francis William Newman 'lately Gardener at Lyndhurst', and George Lindley of whom 'H.E. has no personal knowledge'. James Rennie also made enquiries[156]—and so did Ludwig Leichhardt who had some powerful support. When Robertson's end 'became perfectly evident', William Macarthur,

with the concurrence of several other members of the 'Garden Committee' recommended Dr Leichart to apply for the situation when it should become vacant, promising him our support in case as on a previous occasion the Governor should devolve the choice of a successor upon the Committee of the Botanic Garden.[157]

Macarthur advised Sir William Hooker, Director of Kew, that the young Prussian was diffident, and by 15 July, the governor understood that Leichhardt 'was no longer a candidate for the office'. Leichhardt feared that the xenophobic views of the three Macleays would again prevail, that there would be objections to 'his carrying on a foreign correspondence' for the benefit of the Garden and 'of science at large', and that even if successful, the appointment 'might interfere with the Expedition to reach Port Essington upon which he had set his heart'.

Macarthur had in fact written to George Macleay 'to solicit his vote in favor of Dr Leichardt', only to be informed that 'it is not of much consequence how the situation is filled up just now, as my Father has written to Sir William Hooker to recommend a competent person to the Secretary of State …'. George Macleay himself believed that 'as England is at the lead of Botany as well as of other branches of Science' all such situations 'should be reserved for Englishmen'. Macarthur disagreed. As far as Leichhardt was concerned,

... it would savour of something else besides liberality were we to reject him merely because he is a foreigner for the very doubtful advantage of having it filled by a Secretary of States appointment. If England has attained such remarkable eminence in the Arts and Sciences her advancement is certainly not to be attributed to the exclusion of scientific men merely because they happen to be foreigners ...[158]

Alexander McLeay's action in writing on his own account and without reference to the committee as a whole, had aggrieved Macarthur and his supporters, for they believed 'Dr Leichardt to be a person more eminently well qualified for the situation than we are likely again to meet with. He is something far superior to a mere Botanical Collector'.[159]

In December 1844 Hooker confided to Phillip Parker King that he had heard from Alexander McLeay concerning the

... death of Robertson (I think was his name) at the Sidney Garden & I have been in communication with the Col. Office & urging Lord Stanley to put in a competent person as successor to *Allan Cunningham*. That is the sort of man required, & I shall do my best to have such an one sent out.[160]

For a time it was not clear whether the appointment would be made in Sydney by the Governor on the advice of the committee, or in London by the Secretary of State on the advice of whomsoever he chose. Many shared Leichhardt's view that the Garden was merely 'a place of public amusement and the kitchen garden of Government House, rather than a place for the students of botany and horticulture',[161] and now, in Macarthur's view, there was the opportunity to change this situation to the advancement of colonial science. Macarthur assured Hooker that he was 'authorized in saying that both Dr Nicholson & the Revd W.B. Clarke also members of the Committee will concur with me soliciting Dr Ls appointment'.[162] Macarthur believed that 'this most intelligent and unfriended young man' was opposed only by the Macleays, and so Hooker could expect to hear also from Captain Phillip Parker King and his friend, John Carne Bidwill, an accomplished botanist, who had 'arrived a few days since from New Zealand' and who had already 'suggested that Dr Leichardt should at once write to you to solicit your interest in his favor with Lord Stanley for the appointment ...'.

On 20 July 1844 Alexander McLeay advised Governor Gipps that the committee had met to nominate 'a proper Person' from the six applicants for the position, which 'necessarily requires considerable proficiency in the Science of Botany'. While hoping that a fully qualified person 'will speedily be sent from England' the committee recommended that James Kidd be appointed in an acting capacity at Robertson's salary of £140 year. The committee had agreed that 'as a temporary arrangement, no better appointment can be made'. It was further recommended that Edward Woodhart, a gardener 'known to have had considerable experience in the supervision of Convicts' should be appointed overseer. Gipps approved and Kidd was appointed as from 1 August.[163] The Macleay faction had won, the Macarthur faction had lost, and a few days later Ludwig Leichhardt sailed for Moreton Bay to organize his northern expedition.

Kidd expressed his gratitude, pointing out that he would do his best, but owing to the 'exhausted state of the Gardens, there can be little satisfaction given at present'. He also sought the Reverend W.B. Clarke's advice on arranging Robertson's papers and conducting correspondence.

The Acting Superintendent travelled to the Paterson and Hunter Rivers to collect for Sir William Hooker, attended to the repair of 'Gates &c.', won an argument with council authorities when 'the usual supply of Water from the Race Course' was refused except on payment of 'Sixpence per Cask', and sought regular Boards of Survey to review the state of repair of implements and equipment.[164] In April 1846, Kidd reported receiving from Leichhardt after the Port Essington expedition, 'two hundred kinds of seeds, which have been carefully sown and have yielded new plants'. At the end of July 1847, Kidd listed the seeds and plants despatched and received in exchanges made 1844–47, thereby demonstrating that he had maintained a long-established practice.

There were the usual convict problems. Kidd was charged with ensuring that 'Prisoners belonging to the Gangs' under his superintendence were prevented from 'inter-mixing with the Citizens' during 'the day fixed for the approaching Election of City Councillors'. He reported that the convicts were mustered at seven each evening, then placed in the charge of a watchman until six next morning, and there were further musters 'at different hours of the Night when least expected by them'. The sleeping quarters were, it seems, insecure to say the least. One convict, Martin Heffernan, was actually tried for robbery committed on the Parramatta Road near the Glebe at eleven one night when he was believed to have been snug and secure in the Botanic Gardens barracks, and Kidd was obliged to point out that although there were twenty-four on the establishment, with 'so many absconding the number is seldom more than from 18 to 20'. He suggested that additional rations might help maintain numbers. In mid-1846 matters were not improved by the resignation of Edward Woodhart as overseer, although he was immediately replaced by William Waterman.[165]

Of more professional interest was 'the substance called *Guano* which for some time past in Europe, has attracted much attention as a valuable manure'. It had recently been discovered on the Laurence Rocks, at Portland Bay and the Committee thought experiments might be carried out. Gipps authorized expenditure of £50, but stated that if he were 'rightly informed Guano may be obtained from Lake Macquarie'. Ten tons were ordered from Portland Bay. Three grades were recognized—upper, middle and lower, the last 'supposed to be the best'.[166]

Meanwhile, in London the Colonial Office maintained that no report of the death of Robertson had been received. Sir William Hooker was therefore in an embarrassing situation, being urged privately and frequently to recommend Leichhardt for a position which officially had not been vacated! Actually Hooker had reservations about both Leichhardt and Kidd. In March 1845 he advised Alexander McLeay:

I have a long letter from McArthur strongly recommending *Leickhard*! From what you and Bidwill say I am confident that Kidd is a most unfit person & shall take care to tell Lord Stanley so & having heard a good deal of Leichhart in Paris, where he is well known, & in *many respects* well spoken of, I am equally satisfied *he* is not a suitable person.[167]

Hooker appreciated the problem of upgrading a Botanic Garden in the former penal colony. Clearly the difficulty was to find 'a suitable person ... one who is not a common Gardener, but a man of respectable sphere of Society, & not likely to give way to that horrid vice of tippling'. Apart from the scarcity of such men, there were other obstacles. First, the low salary of £200 was unattractive, and second, 'the convict labourers of the Gardens & some other arrangements there, are such as a scientific man & a person of gentlemanly feelings cannot put up with'. Hooker thus reiterated the complaint of Allan Cunningham that he had to perform the duties of a superintendent of convicts as well as those of a superintendent of the Garden.

In September 1845 bewildered officials in the Colonial Office still maintained they 'had no official intelligence of the death of Robertson'.[168] Poor Gipps was in trouble again, and Stanley was owed an explanation. The Governor was not very convincing: 'On looking through my Despatch book, I regret to find that I omitted, in ... July 1844, to report to your Lordship the death of Mr Nasmyth Robertson, the keeper of the Botanic Garden at Sydney'. Gipps advised that Robertson, 'not a scientific Botanist, but simply a good practical Gardener', had been succeeded by Kidd, a man of similar attainments.

The Governor was aware of a decline in the standard of the Garden. In fact, he confessed 'the term '*Botanic Garden*' is now almost a misnomer, since the Garden is scarcely to be looked upon as more than a very agreeable promenade for the inhabitants of, and sojourners in Sydney'.[169] This decline had been matched by a reduction in the annual vote of the Legislative Council from £1214 (in 1838) to £500 (for 1846) which sums 'formerly included a salary of £200 for a Colonial Botanist', but the title had 'fallen gradually into disuse', with a gardener being in charge at £140 year.[170] If economically sound, this was scientifically disastrous.

By now Hooker, thoroughly irritated by the repletion of correspondence and bungling, had waxed uncharitable:

Now it appears that a Mr Kidd is put in, an old school-fellow of our Mr J. Smith (our Curator) & a *Convict*! or who was a convict. I have however individually no reason to find fault with him: he has sent voluntarily 2 excellent Cases of plants & signs himself 'Superintendent'.[171]

Although Hooker felt that the opportunity to upgrade the Sydney Garden had already been lost, the bungling had not yet reached its peak. The forgetful Gipps left Sydney in July 1846, having been the first vice-regal incumbent of the new Government House built on the Government Domain, graced by plants selected by James Kidd. Sir Charles FitzRoy arrived as Governor in August 1846, by which time the new Secretary of State, William E. Gladstone, had penned a despatch regretting 'that the Garden has lost the scientific character which it originally had'. Gladstone urged the new Governor to seek a report from the Executive Council indicating 'how far ... it might be practicable to restore that character to the institution, without a Sacrifice of the purposes of recreation, to which it appears to have been of late wholly devoted'.[172]

FitzRoy put the matter before his Council in January 1847. The Councillors saw

no reason to think that the use of the Botanical Garden as a place of recreation would be in any degree interfered with by the restoration of its Scientific character. Nor do they perceive that anything more is wanting to effect this desirable object than the appointment of a Scientific person ...[173]

This nicely threw the onus back upon the Governor and the Secretary of State. FitzRoy felt that 'a competent person may be found in this Colony'. Such, he believed, was John Carne Bidwill, 'a gentleman of superior qualifications', and one who was 'perfectly competent to conduct a correspondence with the Botanic Societies of other Countries, and to restore the Institution to the Scientific character, which it is desirable it should maintain'.[174] Earl Grey, Gladstone's successor, was advised accordingly, and on 21 August 1847 Kidd was informed that as from 1 September he would resume as overseer at five shillings a day, since on that date, Bidwill would become Director of the Botanic Gardens.[175] Kidd was to remain overseer for almost twenty more years.

John Carne Bidwill was actually appointed 'Government Botanist and Director of the Botanic Gardens', a title which suggested that a new scientific era for the institution was assured. William Macarthur certainly believed so, as he indicated to Richard West Nash, advocate-general of Western Australia and a keen agriculturist:

Our Botanic Garden here is in a fair way to be completely remodelled & reformed—an intimate friend Mr J. C. Bidwill has recently been appointed 'Director of the B. Garden' with a salary of £*300*. He is, besides being an excellent botanist, & man of general science, a very skilful horticulturalist—perfectly devoted to gardening in almost all of its branches. He will be most happy to open a correspondence with Western Australia ...[176]

Governor FitzRoy must have felt some satisfaction in resolving a rather difficult situation from within the colony, and we can only imagine his astonishment, if not alarm, on receiving Earl Grey's despatch advising that he had 'appointed Mr Charles Moore to be Superintendent of the Botanic Garden at Sydney' on the strongest recommendation of Dr John Lindley, the widely acclaimed botanical and horticultural author and professor.[177] Grey, too, must have been astonished to receive FitzRoy's advice concerning Bidwill, and he testily told the Governor '... it is sufficient for me to refer you to my Dispatch of the 10th of July, in which you were informed that I had selected Mr C. Moore'.[178]

It seems probable that Earl Grey, aware of the pressures being exerted upon Sir William Hooker, decided to seek an independent recommendation. Accordingly, he requested Professor Lindley 'to select a person to fill the office of Director of the Botanic Gardens in Sydney'.[179] From his own knowledge, and with the support of Professor John Henslow,[180] Lindley recommended Moore, who was promptly appointed.

Young Mr Moore thanked the Secretary of State, and went to Kew to tell Sir William Hooker of his good fortune. Hooker, exhausted by the barrage of requests he had received, yet mindful of the great advantages to Kew if the Sydney Garden were upgraded, was not in a congratulatory mood: 'I can scarcely congratulate you upon that, in as much as the appointment has been in my hands for the last three years, and you appear to have stepped in and taken it away from me'.[181] Nevertheless, Hooker wished Moore well, and gave him letters of introduction to Alexander McLeay, John Carne Bidwill, 'and other gentlemen'. The letter to Bidwill explained that his appointment had not been known at the time Grey gave the position to Moore.

Sir William Macarthur, youngest son of John Macarthur, a foundation member of the Botanic Gardens Sub-Committee, and an admirer of Ludwig Leichhardt.
(Mitchell Library, Sydney)

At least one committee member revealed his irritation. 'What a pretty mess they have made in the appointment of Mr Moore', declared William Macarthur in a letter to Hooker, and this in spite of 'Mr Bidwill's remarkable qualifications & the assurance that he was to have the appointment'.[182] Macarthur also confessed his disappointment to Professor Lindley, not because he wished to attack Lindley or his nominee, but because of the high opinion he held of Bidwill, who it seemed, could have quickly rectified the lamentable fact that 'the Garden though large & containing numerous fine specimens has long been a disgrace to the Government'.[183]

In any case, the decision was final. Just before Christmas 1847 the embarrassed FitzRoy had to advise Bidwill of the supersession and to express his apologies.[184] To make the task even more difficult, the Legislative Council voted that the new Director should receive £300 a year.[185]

John Bidwill, who had charge of the Gardens between 1 September 1847 and 1 February 1848, accepted the decision with commendable forbearance and good grace, to take an appointment as Commissioner of Crown Lands, Wide Bay—a job which shortly cost him his life. The respected botanical correspondent of P. P. King and William Macarthur, Bidwill laid out a botanic garden at Tinana, Maryborough, but later became lost in the bush for eight days while surveying a road to Moreton Bay, and died from exposure on 16 March 1853, aged thirty-eight. Ironically, FitzRoy authorized Moore to transfer Bidwill's Tinana plants to Sydney.[186]

Just five years earlier, in March 1848, Ludwig Leichhardt, then thirty-four, had led his last expedition west from the Condamine River, to perish on a proposed transcontinental journey, perhaps a thousand miles from the Gardens he had enriched and had hoped to administer. In July Leichhardt's old antagonist, Alexander McLeay, then eighty-one, having expressed 'his most affectionate dying remembrances to his dear friend Robert Brown',[187] died at Elizabeth Bay House, not far from the Gardens he had supported and had helped to administer.

Ahead of the liberal-minded William Macarthur still lay the award of a knighthood, and nearly thirty-five more years of active and useful life as manager of estates at Camden, commissioner at international exhibitions and member of the Legislative Council and other worthy bodies including the trustees of the Australian Museum and of the Free Public Library, and the Senate of the University of Sydney.

Chapter 5

CRISIS AND CONSOLIDATION 1848–96

I know there was not a single plant labelled in that Garden, when I came to it; I know there was not a single effort at any arrangement in that Garden, when I came to it; and indeed, when I came to it, I am sorry to say, it was no credit to the Colony.

CHARLES MOORE, 1855[1]

Charles Moore had the understandable anxiety of a twenty-seven-year old who had resigned one position in England for a much more responsible one on the other side of the world. In July 1847 he sought assurances about preparation time for the voyage, commencement of salary, conditions of employment, to whom he would be responsible, and the matter of accommodation. Earl Grey avoided some of the queries, but advised that a free passage would be recommended, while provision of a house was in doubt. It was considered that six to eight weeks would enable adequate preparations to be made, and the young botanist was told to place himself 'under the directions of the Governor on arriving at Sydney; but the practical arrangements connected with the Botanical Garden are under the Committee of Management'.[2]

Joining the steamer *Medway* on 18 September 1847, Moore found that his fellow passengers included the first Bishop of Newcastle, the Right Reverend William Tyrrell, who promptly requested him 'to read an essay on the utility of Botany for the benefit of those on board'.

Moore arrived in Port Jackson in mid-January 1848 under a cloud which was not of his own making. The rejection of Ludwig Leichhardt and the ousting of John Carne Bidwill had stirred anew the whole issue of colonial appointments. William Macarthur described the mood of the colonial legislators for Dr John Lindley's benefit:

We had managed to get the members of our Legislative Council to consent to an increase in the Salary to the Head of the Garden from £200 per ann. to £300 upon the understanding from the local government that Mr Bidwill should be appointed to

Charles Moore, Director of the Botanic Gardens, 1848–96.
(Mitchell Library, Sydney)

it—They almost to a man declared they would not vote this increase if a person in England was to be put into the situation, and you may imagine that they are not well pleased to find that after all the situation has been filled up from thence—

There was no intention to question Lindley's judgement, but the Sydney Garden was considered to be 'a disgrace', and Bidwill was believed to be the best man to restore it, because of 'the vast extent of his information upon a variety of subjects' and 'his zeal in the practice of experimental horticulture & in the pursuit of science in general'. Macarthur continued:

I have no doubt that we shall find Mr Moore an efficient officer, but we cannot expect that he should prove at all equal to the gentleman whom he unconsciously supplanted—You are acquainted with Mr Bidwill ... In his removal I assure you we feel as if we had our right hand lopped off ... I now fear our Council will have their backs set up & Mr Moore will have a difficult card to play ... These acts of official interference make colonists at a distance very sore & disposed to kick when occasion offers ...[3]

The main issue was not where an appointee was born, but where, and by whom, an appointment was made. Bidwill himself was born in Exeter, but for the Colonial legislators, mindful of their growing independence, he had the unquestionable advantage of nearly a decade of colonial experience. Even the Prussian-born Leichhardt had spent some six years studying and exploring his adopted country. But what of Charles Moore?

Like several of his predecessors, Moore was Scottish-born, a native of Dundee, where the family name had been Muir. The family moved to Ireland where as boys Charles and his elder brother David studied in the Botanic Gardens of Trinity College, Dublin, under the supervision of the first curator, Dr James Townsend Mackay. David was appointed botanist to the Ordnance Survey of Ireland in 1834, and was shortly joined by Charles. In 1838, David became Director of the Glasnevin Botanic Gardens, Dublin, while Charles, after moving to London, gained further horticultural experience at Regent's Park and at Kew. Having 'stood a most creditable examination' before Professor J.S. Henslow and being in high favour with Dr Lindley, he was recommended to Earl Grey for the Sydney post.

On 27 January 1848 the Colonial Secretary sent Bidwill the final painful advice in typically uncompromising terms: 'Mr Moore, who has now arrived in the Colony, will immediately take charge of the Garden, superseding you in the Situation of Director of the same'.[4]

Charles Moore entered on duty on 1 February 1848 and found much required attention. There was a Gardens residence which required repairs to the extent of some £30; gates and fences also required repair, but fortunately £100 had just been set aside for this work. Permission had to be sought to take action which incurred expense, and the new Director was given a basic lesson in bureaucracy: 'you will confine yourself in each communication to one distinct subject'.[5]

With commendable celerity, a detailed memorandum of duties for Moore's guidance was issued by the Colonial Secretary early in March. The shortcomings of the past were to be remedied without delay:

It is intended that the Garden shall become a place where the science of Botany and Horticulture may henceforth be studied upon the most improved system. At the same time, the Garden is to combine with these objects a pleasant place of resort to the inhabitants of Sydney.

Moore was to undertake the general superintendence of this Garden with the aid and advice of the Committee of Management; to correspond with other institutions and private individuals throughout the world to effect exchanges of seeds and plants 'likely, in a scientific or economical point of view, to be beneficial to the countries they inhabit, or to this Colony'; the plants in the Garden were to be labelled, and an accurate catalogue compiled; a Code of Rules, relating to the distribution of plants and seeds, was to be prepared noting that 'as a general principle ... no seeds and plants shall, in future, be distributed from the Garden, which are procurable from the private Nursery Gardens'. Furthermore, an annual report was to be submitted; collecting excursions were to be undertaken 'with the sanction of the Governor'; an annual 'course of Lectures on the elementary principles of Botany and Vegetable Physiology' was to be delivered. The Garden would in fact become 'a school of Horticulture, where the best modern systems of cultivation may be exhibited to the public'. In performing these duties the Director was 'to submit to the Committee, for their advice and assistance, all matters of importance affecting the management of the Garden'.[6] As a foundation member of that committee, and Fellow of the Linnean Society of London, it is likely that Edward Deas Thomson, the Colonial Secretary, composed these instructions himself. He sent a copy of the memorandum to the committee for information, pointing out that it had been given to Moore 'for his guidance in the performance of his duties'.[7]

The Director's residence, photographed about 1855. Charles Moore stands in the garden, his niece Isabel and wife Elizabeth are on the balcony. The building was demolished in 1875.
(Royal Botanic Gardens Library)

Further guidance was readily available from the state of the Garden itself. Moore inherited an institution which had generally run down after years of financial restrictions, and which had suffered from a lack of clear scientific, economic and recreational policies and from frequent changes in superintendents. To make matters worse, it was now over seven years since the transportation of convicts to New South Wales had officially ceased, resulting in a 'considerable diminution' in the amount of free labour permissible 'within the Estimates'. Consequently, Moore found that 'the Garden was necessarily in bad condition, the Borders were overgrown with weeds, and the Walks out of repair …'.[8] There was also the problem of the Crown reserves adjacent to the Botanic Gardens. Was the Director also to assume responsibility for directing the ordinary operations of the Inner and Outer Domain and Hyde Park, he asked a fortnight after taking office. Eight months later he was advised that 'it will be your duty to direct the ordinary works in the Inner and Outer Domains and that the management of Hyde Park has been undertaken by a Committee approved by His Excellency'.[9]

Charles Moore's design for a plant label proposed in 1848 was the subject of consideration and annotation by the Colonial Secretary, Edward Deas Thomson, and Governor FitzRoy.
(NSW State Archives)

Moore dutifully wrote to the Committee of Management seeking advice on rebuilding the arbour and submitting for approval specimens of labels for the established plants.[10] In April 1848 he suggested some guidelines for the distribution and exchange of plants and seeds. There should be free exchange 'with all other Gardens of a similar description' and with other public establishments able 'to make an equitable return of a like nature'; when similar returns were guaranteed, exchanges might also be made with private individuals in Australia and overseas; plants and seeds of economic value 'may be distributed to all persons making application', and a register was to be kept. The committee 'entirely concurred', and Governor FitzRoy approved.[11] In the following month, Moore's proposed stringent regulations for the Garden were also warmly approved and immediately gazetted:

1. It will be in the discretion of the Director of the Garden, or of those acting under his orders, to refuse to admit all persons of reputed bad character; all persons who are not cleanly and decently dressed; and all young persons not accompanied by some respectable adult.

2. Smoking will be strictly prohibited in the Garden. Any person offending against this Regulation will be removed from the Garden.
3. Any person found taking or injuring plants, flowers, or fruit, will be summarily removed from the Garden, or proceeded against by law according to the nature and extent of the offence committed.
4. The Garden will not be open to the public on *Sundays* until after the hour of 1 o'clock, p.m.[12]

During his first year, Moore repaired 'nearly all the Walks in the Upper Garden' (now known as the Middle Garden) and constructed '600 yards of a new Walk, 9 feet wide, in the Lower Garden' with over 900 cart-loads of gravel. A further '300 cart-loads of stone brought from the Old Government House and elsewhere' were used to erect '200 yards of Sea Wall' around Farm Cove, and so was begun the structure which was later nicely dubbed 'the Thirty Years' Wall'.

In April 1848 the Director sought the committee's opinion

relative to the System which should be adopted in the arrangement of the plants ... the Natural Method is altogether the most comprehensive, and most complete; yet the Linnean or Artificial System, presents many advantages to the young botanical students ... and is therefore worth consideration, before deciding in favour of the former.[13]

Moore tactfully used both systems when arranging certain plots, and 'the Committee highly approved'.[14]

The two great problems in carrying out plans for development, were of course, labour and cost. In preparing estimates for 1847, James Kidd had reckoned on having twenty-four convicts as a labour force, requiring an annual outlay of £2-2-9 each for clothes, in addition to daily rations, which he carefully calculated at 24 × 365 = 8760 @ 3d = £109-10-0. Rations for the two Gardens horses however, were calculated at $7\frac{1}{2}$d day, thus: 2 × 730 = 730 @ $7\frac{1}{2}$d = £22-16-3, but '2 setts shoes monthly @ 4/-' would cost an additional £4-6-0.[15]

A week before Moore assumed office, the Colonial Secretary advised:

now that Prisoners cannot be obtained I am directed by His Excellency ... to inform you that the Sum of £220 is voted for Provisions and Clothing for twenty four Convicts for the Establishment for ... 1848 and that this Sum may be applied to the employment of free labor it being understood that the amount £220 must not be exceeded in the year.[16]

On taking up duty, Moore complained of the inadequacy of the sum allowed for free labour, and he was asked to submit a renewed proposal on the subject. Within three weeks the Governor had approved the employment of 'nine men in the Botanic Garden ... at the rate of three shillings per day each together with another at the same rate as Propagator'.[17] The days of employing large convict gangs for moving masses of sandstone and for clearing and filling operations were over.

Botany and bookkeeping aside, Moore, like his predecessors, soon appreciated some of the many extraneous problems associated with administering the Garden, and he must have enjoyed going away briefly on his first collecting trips to the Illawarra and Hunter River. Doubtless encouraged by the fact that the Lower Garden had been made available to the Australasian Botanic and Horticultural Society on 11

October 1848, the band of the 11th Regiment sought permission to play there later in the month, and thereafter once a week. The committee promptly dealt with this proposal, thereby sparing the new Director from appearing unduly officious. Committee Secretary, the Reverend George Edward Weaver Turner, Rector of St Anne's, Ryde, advised the Colonial Secretary that 'injury must necessarily accrue to the Garden, if, contrary to the newly-established Regulations, all persons were to be indiscriminately admitted who usually attend the Band'. Clearly the Domain was the place for such recitals, for there, all who so wished could attend without threat to 'the management and preservation of the Botanic Garden'. It was hoped that His Excellency and Major General Edward Wynyard would understand.[18] The band played elsewhere.

In the meantime, as William Macarthur predicted, the colonial legislators and administrators felt 'disposed to kick'. Moore's salary was to be reduced to £200, and while it stood at the higher rate of £300, he was required to pay £91-9-6 for his 'free' passage.[19] In the Council, Stuart Alexander Donaldson (soon to become first Premier of NSW under responsible government) endeavoured to have the vote for the Gardens so reduced that the new Director would have been virtually starved out in his first year.

Accordingly, Moore rejoiced in his candid friendship with William Macarthur, who 'during a three hours' conversation ... in the Botanic Garden' outlined 'everything that had taken place', thereby enabling the ostracized Director to play his difficult card with confidence and effect. Macarthur also kept Sir William Hooker informed:

Mr Moore is getting on pretty well at the Sydney Gardens—that is he has certainly effected considerable improvements by removing old and useless specimens, and by causing the grounds to be neatly kept—I do not perceive that he has hitherto made any collections of indigenous plants—in fact he has much to learn before he becomes useful as a collector—He is improving however and I think would soon become very serviceable if he would but drop a little of his self conceit—I have uniformly found him most civil and obliging.

The Legislative Council, as was to be expected, cut the salary down from £300 to £200—so much for the Downing Street meddling ...[20]

Clearly Mr Moore had managed notwithstanding considerable prejudice, and in January 1849, when he presented his first Report, there was ample cause for satisfaction. Despite the shortage of labour much landscaping had been done; walls, walks and fences had received attention; many new and desirable plants had been introduced and material had been sent in exchange; and the new Rules had 'had the desired effect of preserving order and decorum, without in any way interfering with the liberty of persons visiting the Garden'. It followed that 'the character of the Garden has been raised sufficiently in the estimation of the Public, to give an assurance that within its limits all the privacy and retirement of a rural walk may now be enjoyed without fear of interruption by rude or disorderly persons'. Mindful of his instructions, the Director had aimed to make 'the Institution an object of interest and importance, as well for the study of Botany and Horticulture, as for the pleasure and recreation of the Public'. One area in the Lower Garden had been 'remodelled' to illustrate 'the natural orders included in the class "Monochlamydeae"'[21] and another in the 'Upper' (i.e. Middle) Garden had been prepared 'for a collection of such Plants as are used in the Arts, Manufactures, Medicine, and for Domestic Purposes'. A catalogue was in preparation, and specimens growing in the Gardens were being labelled 'shewing the Natural Order, Scientific Name and Authority, English Name and Native Country of each Plant', a practice which had 'given very general satisfaction',[22] and which was continued. Authors have been omitted in recent years, and the use of blue labels now enables Australian species to be readily distinguished.

Moore paid tribute to supporters whose donations had been especially welcome— 'the Messrs. Macarthur of Camden', George Macleay, Dr Charles Nicholson, Reverend G.E.W. Turner, Sir Thomas Mitchell and Dr George Bennett, who donated seeds of the *Clianthus* 'found in the Desert by Mr Sturt'. A final tribute was paid to 'the Government and ... the Gentlemen constituting the Committee of Management' for their 'uniform kindness and support'.[23] Governor FitzRoy was delighted: 'I think this Report does Mr Moore much credit. He sh^d be made aware that I consider the Zeal & Efficiency he has displayed in getting the Gardens into order as deserving of the commendation of the Government ...'.[24]

Early in his second year, Moore obtained permission to undertake seed and plant collecting excursions to Camden and Appin, Port Stephens and the Hunter River. Material was also received from William Carron, whom Moore had recommended to accompany Edmund Kennedy's ill-fated Cape York expedition as botanist in 1848, and from Lieutenant-Governor Edward John Eyre of New Zealand, also associated with Australian exploration. In June 1849, Carron, then back in Sydney at Shepherd's Darling Nursery, wrote to the Committee seeking a position in 'a Hortus Siccus in connection with the botanic Garden and Museum—should such be established ...' but it was not until September 1859 that Carron secured a position at the Gardens at the rate of 6/6 per day. Appointed 'Collector' in 1866, he remained until the end of 1875.[25]

Moore's early associations continued, literally, to bear fruit, and welcome donations of seeds were received from the Royal Botanic Garden, Regent's Park, London and from the Royal Botanic Gardens, Glasnevin, Dublin where his brother was still Director. John Carne Bidwill sent seeds 'and some fine plants of a species of Cymbidium' from Wide Bay, and further material was received from other donors in Britain, from the other Australian colonies, the Pacific and Asia. Work continued on the walkways, and the long walk encircling the lower garden was completed; 'the old and comparatively useless brick pit' was demolished and replaced by a large plant house for the propagation of plants, and raising seedlings. Free labourers converted the old convict barracks into storerooms and other facilities.[26]

Moore sought permission to remove such materials 'from the old Barrack wall and Treasury, as are suitable for the erection of the Sea Wall in the lower Botanic Garden, and for making a Drain from the Hospital wall, through the Outer Domain ...' but the Governor disapproved.[27] In his Report for 1849, Moore stressed the need for an adequate water supply, which in dry seasons was 'limited to a few casks per diem, from the Hyde Park fountain'. Already 'some rare and beautiful plants' had been lost, and there was the need for 'a direct channel for the conveyance of water to the garden'.

Work proceeded on 'laying out *Botanical arrangements* of plants, according to their *natural* and *sexual* affinities'. William Sharp Macleay and Dr Archibald Shanks of the

*The tendency of the Botanic Gardens Creek to silt up is shown in this photograph, attributed to
Alexander Brodie, about 1870.
(Historic Photograph Collection, Macleay Museum, University of Sydney)*

Committee of Management inspected this work and 'were much gratified ... by the progress made' and indeed, 'by the neatness and efficiency which were apparent in every department of the Garden which they inspected'.[28] Moore paid a tribute to Charles Fraser and the Cunninghams by declaring that 'the general good taste displayed in the original design of the garden, has rendered it unnecessary to effect any important alterations in this respect ...'.

Once again Charles Moore's endeavours met with official approbation. The Committee Secretary, the Reverend George Turner, relayed the Committee's approval to the Colonial Secretary, who advised that he in turn had informed the Governor. The committee had intimated its 'entire satisfaction with the proceedings of the Director of the Botanic Garden' and His Excellency was 'much gratified to observe that you coincide with Him in the opinion He has formed of Mr Moore's merits'.[29]

Mr Moore's merits were more widely appreciated. Bereft by the loss of the naval officer to whom she had been betrothed, Elizabeth Bennett Edwards is said to have met the Director during one of her sad meditations at Mrs Macquarie's Chair. On 7 July 1849 they were married at St Anne's Church, Ryde. Appropriately, the Reverend George Turner officiated, and the guests included two other well-known naturalist-clergymen, the Reverend W.B. Clarke of St Thomas's, North Sydney and the Reverend James Walker of St Luke's, Liverpool.

In 1850 Moore asked where he might deliver lectures on botany, and the Colonial Architect thought that 'the Sheds in the Garden' might be modified for £45. It was an improvised amenity: 'By taking the bricks out of an old drain I put up two walls to the old barrack', said Moore in 1855, and 'then got a sum of money from the Colonial Architect, to roof and plaster the place'. When the first lectures were given in 1851 in order 'to introduce a taste for the science of Botany', Moore was pleased to note that they 'met with greater success than I anticipated, the large attendance of persons of both sexes, and the interest which they appeared to take in the subject was to me satisfactory and encouraging'.[30] There was no Faculty of Science in the University of Sydney until 1882, and no School of Botany until 1913. Medical students attended Moore's lectures, and between 1880 and 1882 young Mr J.H. Maiden was in the audience.[31]

Early in 1850 Moore was given permission to visit the Blue Mountains and Myall Lakes, and later in the year, to join HMS *Havannah* on a voyage to the South Seas, 'for the purpose of obtaining a collection of the vegetable productions of those Islands'.[32] He took 'glazed cases for plants, and materials for collecting seeds and specimens'.[33]

Doubtless Moore noted with pleasure that the vote for 1850 included £100 'to complete the Sea Wall, and the Salt Water Pond at the Western side of the Lower Garden'—a sum which unfortunately proved totally inadequate partly owing to the deep foundations required; £100 'for making Drains and fresh Plantations in the Outer Domain'; and £140 'for improving the Inner Domain'.[34] Not quite so pleasing was Governor FitzRoy's decision, made while Moore was in the South Seas, to permit the Church authorities 'to draw from the quarry in the outer Domain, the stone required for enlarging St Mary's Cathedral provided care be taken that no injury be done to the Domain in the course of the operations'.[35] A similar situation arose early in 1852 when Moore was asked to report on a proposal by the Sydney Corporation 'to open that portion of the Domain north of Sir Richard Bourke's Statue for the purpose of removing the Iron Stone Gravel to repair the City Roadways'. Moore refused to entertain the idea, and the Governor advised the Town Clerk accordingly.[36]

By this time, Moore's status had changed significantly. On 15 November 1851 the Committee of Management

having taken into consideration its position, with respect to the supervision of the Botanic Garden, thinks that, unless it be deemed expedient that some definite instructions be given by the Government to authorize its direct interference, the Committee should be relieved from the present reference to it, which can scarcely be attended with any beneficial results.[37]

Contrary to expectations, FitzRoy was loath to grant extended powers but he took part of the advice offered, and promptly relieved 'the Committee from any supervision of the Botanic Garden', adding somewhat tepid thanks for services rendered.[38] Moore thereby became virtually a free agent, responsible only to the Governor, through the Colonial Secretary. Naturally the severance of the committee's association with the Gardens did not allay the resentment which, stirred by certain private nurserymen, grew threateningly.

Meanwhile, another crisis had arisen. On 15 May 1851 the *Sydney Morning Herald* reported Edward Hammond Hargraves's gold discovery near Bathurst. Five days later, Moore reported that men were leaving employment in the Gardens and Outer Domain 'for the purpose of proceeding to Bathurst, and ... others intend following them, unless they receive higher wages ...'.[39] Moore was empowered to increase wages to retain some of the men but of course, the overall estimates could not be exceeded. In January 1852 the following rates were approved on Moore's recommendation:

For the Botanic Garden		For the Outer Domain	
1 Propagator	at 4/6 per diem.	2 Workmen	at 4/– per diem.[40]
7 Workmen	at 4/– per diem.		
2 Carters	at 4/– per diem.		

Moore's own salary was at last restored to £300 year.[41]

About this time the Annual Report on the Botanic Gardens assumed a new and more detailed form. Moore submitted his Report for 1850 in the usual way, referring to progress on the sea wall and on the plots to be arranged in systematic order, to the receipt and despatch of seeds and plants in general terms, and so on. However, the Governor returned the report, pointing out

that it would have been much more satisfactory, had there been appended ... a list of all the plants and seeds which have been added to the collection already existing in the Colony, since the last report, and also of the contributions made to other Institutions and Countries, by way of return ...

Moore was asked to furnish the information and the final Report for 1850 was voluminous indeed, and certainly much more useful, historically and botanically.[42] Subsequent Reports followed the same pattern.

Disdainful, or simply oblivious of the jealousy and resentment mounting against him, Moore worked through the early years of the Golden Decade with vigour and effect. There were ample matters for attention, both routine duties and innovations: the receipt and despatch of botanical material; the acclimatization of desirable ex-

Charles Moore's long-term reclamation project in Farm Cove added about 12 acres to the Lower Garden.
In 1872, silt barges were still in use.
(National Library of Australia, Canberra)

otics; the progressive reclamation of the inter-tidal mud flats at the head of Farm
Cove; the consolidation and extension of the sea wall around the realigned cove; the
maintenance of existing pathways and the construction of new ones; the turfing of
lawns and superseded paths; the formulation of plans for a reliable water supply;
the trenching and manuring of cultivations, and generally improving the meagre
layer of indifferent soil on the sandstone bedrock by means of silt dredged from the
harbour, the sweepings of the well-manured city streets, refuse from the slaughter-
house (presumably 'blood and bone' in its most literal form) and other material de-
fined simply as 'rubbish'.

In 1851 plantings of a considerable number of rare trees and shrubs were made
according to A. P. de Candolle's four natural divisions: Thalamiflorae, Calcyciflorae,
Corolliflorae and Monochlamydeae, 'divisions so natural, that the arrangement cannot fail to be understood and appreciated by persons at all conversant with the
science'.[43] Another innovation was the establishment of a plot 'with a selection of
plants used for medicinal purposes'. Moore hoped that these carefully labelled plants
would 'be instructive as well to general visitors as to medical students'. He also 'humbly and most respectfully' begged that 'a glass roofed conservatory' be built for tropical plants, and reported progress with the introduction of British fodder grasses. He
felt that the problem of an adequate water supply might be solved by laying down a
water pipe from Macquarie Street into the Upper Garden. Two years later, the sum
of £150 was allocated accordingly.

The goldrushes continued to lure away the labour force, and Moore lost 'nearly all the workmen acquainted with the duties of the Garden'. Regrettably, they had been 'replaced by men in most instances totally unacquainted with Garden labor'.[44]

Early in 1852 Moore's request for a boat was declined, but a vote of £50 was granted for 'the formation of a Public Botanical Library'. Twenty-six works were procured, representing such celebrated authors as Robert Brown, John Lindley, William Jackson Hooker, John C. Loudon, Carl S. Kunth, Augustin P. de Candolle, Kurt Sprengel and Stephen L. Endlicher.[45]

Moore's maintenance of uncompromising vigilance over the grounds entrusted to his care, while praiseworthy, was sometimes considered rather too dedicated. Early in August 1852, before leaving on a Spring excursion to the Castlereagh River and Warrumbungle Mountains, the Director complained that the large influx of Sunday visitors was detrimental to the welfare of the Garden. However, Governor FitzRoy considered

you somewhat overrate the damage done by the visitors to the Botanic Gardens on Sundays as (visiting them almost every Sunday himself) His Excellency can take upon Himself to say that He has never seen a single instance of misconduct on the part of the Visitors or any inclination to meddle with or injure the Trees or Plants ...[46]

Nevertheless, His Excellency saw no reason why two or three constables could not be on duty during busy Sundays if for no other reason than to allay the Director's fears for the safety of the plants he tried so hard to protect. The sum of £40-15-0 was to be made available 'for the execution of Diagrams' to render the annual botanical lectures more illuminating. A further proposal was 'to include into one Department, the Botanic Garden, the Outer and Inner Domain and Hyde Park' at a cost of £365-16-0, a sum which would be included in the estimates for 1853.[47]

In the spring of 1853 Moore travelled

through the Northern districts of the Colony ... with the triple object of reporting upon the plants left by the late Mr Bidwell [*sic*] at Wide Bay, selecting Specimens of Timber for the Paris Universal Exhibition, and collecting Seeds and Plants for the general purpose of this establishment ...[48]

The ninety-two timber specimens from Wide Bay and Moreton Bay, duly 'accompanied by dried specimens to verify their scientific names' were to earn their collector a silver medal.[49] It was decided to advise Bidwill's executors that it did not 'appear ... expedient ... to incur the Expense of purchasing and preserving the collection of plants of the deceased gentleman'.[50]

In 1853 and 1854 Moore investigated the problem of sand drifts at Newcastle. As a result of fencing and cultivating, 'the ground which was last spring covered with a loose shifting sand, is now producing a thick luxuriant vegetation ...'.[51] Within the Gardens during these years, Moore established some plants collected on his northern trip, and constructed 'a pond lined with brick and Roman cement' for growing aquatics and storing water in the Upper (Middle) Garden. He noted that in the Lower Garden, 'the only place in which fresh water can at present be collected, is that in which the monument to Cunningham is placed', but the pond was apt to dry up during the summer, to the 'great inconvenience of visitors'. Moore suggested the

*The rustic bridges spanning Botanic Gardens Creek attracted many photographers, including Charles Kerry,
who took this view in 1885.*
(Royal Botanic Gardens Library)

laying of pipes and the construction of a tank to serve the needs of plants and humans. Some 'neat glass roofed brick pits' were built for the cultivation of small and tender plants and among noteworthy donations was some sugar-cane seed received through the kindness of the Colonial Secretary, Edward Deas Thomson.

In mid-1854 Moore reported the genesis of a herbarium:

With the view of enabling the Public more easily to become acquainted with the names of plants, as well as to afford a ready means of reference, a Collection of Specimens of plants, principally indigenous to this Colony, is in progress of being arranged in books prepared for the purpose. The *Ferns* have already been completed as far as possible, and may now be consulted ...[52]

This raises an interesting point. In July 1852 the Surveyor-General, Sir Thomas Mitchell, advised the Reverend George Turner that he was donating to the Australian Museum

a collection of specimens of sub-tropical plants, collected by me during my last expedition ... into Tropical Australia. These plants have been arranged and described by Dr Lindley, Sir William Hooker, Mr Bentham, and Professor De Vries. The specimens are now fit to occupy a place in the museum, and you will perceive that great care is necessary to keep the labels along with the specimens to which they respectively refer. In delivering over to the Curator ... this collection, the proceeds of much personal trouble, but rendered valuable only by the liberal attention bestowed on the plants by the men of science in Europe, I would ... observe that I consider scientific arrangement the first step toward the cultivation and domestication of the indigenous plants and shrubs of their country, a duty one owes to his Creator and to himself; and I believe these natural productions are full of promise, for culinary and commercial purposes.

The Surveyor-General also expressed hope for the establishment of a school of 'Economic Botany in these regions, where bountiful nature seems to await the industrious hand of man'.[53]

These specimens were presented in four packets containing 160 species, with lists of contents which still exist,[54] but they did not pass into sympathetic hands. As a botanist, Turner doubtless appreciated their significance, and apparently either he, or one of his successors transferred the specimens to the Gardens in Moore's time, but the collection was grossly neglected. In 1855 Moore declared that, until 1853, the 'Garden was utterly destitute' of a herbarium, and that when he arrived 'there was not a single specimen'.[55] As early collections customarily went to Kew, the British Museum or to other institutions or private collections in Europe, Moore's claim would have been perfectly true as long as the Mitchell specimens remained in the Australian Museum. Yet, despite Moore's establishment of a herbarium, his successor, J.H. Maiden, found only a dozen portfolios, each with 150 pages, comprising the collection when he assumed the Directorship in 1896.[56] George Harwood, who had joined the Gardens staff in 1873, told Maiden of a 'Cunningham collection' which had been stored in a Gardens cottage vacated by Moore in 1875. Maiden finally found these specimens in an old seed store and they proved to be the remnants of the collection of which Mitchell had been so proud half a century earlier.[57] The full story of this minor botanical disaster and Moore's association with it, if any, will probably never be known.

Despite indications that the Director was performing his duties with commendable energy amid widespread approbation, a crisis was brewing. In accordance with his instructions, he had travelled within the colony and beyond on collecting expeditions. Further, he had entrusted seeds and plants to ships' captains, 'Medical Superintendents of Emigrant Vessels', and other 'persons returning to Europe', and waited for 'some equitable return'. His antagonists considered that he waited too long for too little of what was too common. The desiderata sought by overseas gardeners, nurserymen and botanists still included the favourites of Governor Phillip's time, the obviously unique and Antipodean, species which are today largely protected by law—Christmas Bush, Christmas Bells, Waratah, palms, tree ferns, and so on. With the investigation of the northern rainforests, other species were sought—Silky Oak, Red Cedar, Hoop Pine, Moreton Bay Chestnut and Australian Teak. Even the White Cypress, *Callitris glaucophylla* of the western interior, found its way into the garden of the Duke of Devonshire.

The local nurserymen were by now endeavouring to meet not only local demands, but also the requests of overseas nurserymen, especially in England. As a protective measure, Moore's instructions virtually forbade him to distribute from the Gardens those species which were available from private nurserymen. Complaints about this matter moved the Legislative Council in October 1854 to ask Governor FitzRoy to table 'a Return of all plants, seeds, and specimens distributed from, or received at, the Botanic Gardens ... during the last three years'.[58] William Elyard of the Colonial Secretary's office, advised Moore that this information was required immediately, and the return was 'to be a nominal one, and to distinguish between the plants, seeds and specimens sent to, or received from Public Institutions, and those sent to, or received from Nurserymen or Private Individuals, and to state the number of specimens of each class or species either distributed or received'.[59]

Moore understandably protested that he had already provided this information in the appendices to the very Annual Reports which the Council itself published with its papers. But since identification of individual donors and recipients was required, he was happy to name eleven colonists with whom he had consistently exchanged plants: William Macarthur of Camden; George and William Sharp Macleay; Thomas Ware Smart, later MLA and MLC; Thomas Woolley, a successful merchant who had returned to England in 1850; Reverend G.E.W. Turner of Ryde; Thomas Sutcliffe Mort, the prominent Sydney businessman; Isaac Nichols, 'gentleman', son of the emancipist postmaster; and, most significantly, three prominent nurserymen whom Moore nicely distinguished by 'Mr' instead of 'Esq.' as applied to the others—Thomas William Shepherd, Michael Guilfoyle and John Baptist. The Director was well aware of his instructions, but what means were at his disposal for introducing new species to the colony if he were precluded from sending material in return? He claimed that his 'only desire in effecting exchanges of plants, &c., has been to advance the science of Botany, and to benefit the Colonists by enriching this garden to the greatest possible extent'.[60] If the Governor wished to issue additional instructions, Moore would 'be most willing to attend' to them.

Two of the nurserymen were determined to take the matter further. Shepherd and Guilfoyle[61] considered that Moore's account was 'evidently drawn up with the intention to mislead' and accordingly they submitted a petition to the Council complaining that 'the supply of ... specimens from a Government Institution to the Nurserymen of Foreign Countries is a direct and unfair interference with their trade'. The petitioners claimed that their 'vocation ... is ... calculated ... to develop the resources and capabilities of the country, and to aid largely in the promotion of the moral and social, as well as the physical arts'.[62]

It was charged that Moore wasted public money by sending overseas large numbers of under-packed cases; that plants of equal market value were not being received in return, either from overseas contacts or from local colonists; that local nurserymen should be permitted to buy seeds and cuttings of new discoveries and introductions if they so wished.[63]

Here was the opportunity Moore's antagonists had awaited, if in fact they had not contrived it. Then, in mid-January 1855 a new Governor, Sir William Denison, arrived, with a reputation for having a keen interest in colonial science. On 24 July a Select Committee was formed 'to inquire into and report upon the Management and Conduct of the Botanic Gardens of New South Wales'.[64] The terms of reference could hardly have been wider. The Committee comprised: George Robert Nichols, MLC for Northumberland, chairman, self-styled 'radical reformer' through whom the petition was presented; Charles Cowper, MLC for Durham, who had recently attacked the Government for weak administration; George Macleay, MLC for Murrumbidgee and a keen naturalist; Thomas Barker, MLC, Commissioner for Railways; James Macarthur, MLC for Camden West and a prominent agriculturist and pastoralist; William Macleay, MLC for Lachlan and Lower Darling, and an amateur naturalist; James Wilshire, MLC for Sydney and radical supporter of Cowper and Nichols; Captain Phillip Parker King, FRS, MLC for Gloucester and Macquarie, experienced naturalist and one-time shipmate and friend of Allan Cunningham; Stuart A. Donaldson, MLC for Sydney Hamlets, a 'liberal conservative' who believed in 'spicy opposition'; Daniel Egan, MLC for Monaro, a liberal supporter of Cowper; Captain Edward W. Ward, RE, MLC, Deputy Master of the Mint.

The chairman called for a copy of Moore's instructions and for all correspondence relative to the dissociation of the Committee of Management from the Gardens. Then followed the relentless examinations of the six sessions of the enquiry during August and September 1855. Witnesses called were James Kidd, who had then been at the Gardens for twenty-five years; the Reverend George Turner, secretary of the Committee since 1847; Stuart Donaldson, a proven adversary who had striven to intimidate Moore on his arrival by seeking to have the Gardens vote reduced to £150; and the two petitioners, Thomas W. Shepherd and Michael Guilfoyle.

Probes were made into fields far beyond the supply of plants to colonists and the livelihood of nurserymen. No efforts were spared to embarrass Moore and to condemn his administration. The 591 questions clearly revealed Moore's strengths and weaknesses, his friends and his enemies, and the keen desire of some to curb Moore's authority by saddling him with a new committee. There were questions and wrangles over the desirability of such a committee in the public interest, its size, and whether it should be directive or consultative in function. Attempts were made to show that exchanges had been made on terms botanically and economically unfavourable to the Garden and the colony, and that favouritism had determined the distribution of seeds, plants and cuttings. The Garden accounts were queried, and even issues such as fencing repairs and the construction of plant cases received prominence. Moore's qualifications were seriously questioned, together with his ability to classify and label plants accurately, and aspersions were cast upon his ability to spell. He was accused of having been discourteous to the Committee of Management, and of having neglected to prepare a catalogue of the Garden's holdings. Moore's botany lectures were criticized in their nature and scope, and odious comparisons were made with Bidwill's 'eminent degree' of 'botanical science and general information'.

The unfortunate Director was accused of frequently and arbitrarily dismissing employees, of permitting unseemly dog-hunts during visiting hours, of being so lavish with distributions of plants that private nurserymen were being injured, and of exaggerating the forlorn state of the Gardens when he took office. Moore protested that he had been persecuted, especially by Captain King, ever since he arrived. 'I am either fit for the position I hold,' proclaimed Moore, 'or I am not fit; if I am fit, there is no necessity for a Managing Committee.' He alleged that his salary had been reduced because he 'was represented to be an incompetent man'.

The Reverend George Turner handled the inquisitors well, and supported the Director. He did not know 'a man more competent to manage a Botanic Garden than Mr Moore', who previously had been 'contemptuously treated and discouraged by the Committee'. In Turner's view, Phillip Parker King, James Macarthur, George Macleay and Sir Charles Nicholson had from the outset all been opposed to Moore whom 'they did not speak of ... as a botanist', but 'as one unfit for the situation he held'.

James Kidd answered impartially and honestly, but it was Donaldson who dropped the greatest bombshell. He maintained that Sir William Hooker, 'would not, for one moment, have thought of Mr Moore' as fit to replace Bidwill. Hooker had not known of Bidwill's appointment, and allegedly had said: 'Mr Moore is not, in my opinion, a scientific botanist. I have known him for some time; and he is an excellent practical head-gardener, but not a *botanist*, in the sense in which I understand the term'.[65]

Other factors no less significant were also revealed. One was dissatisfaction over Moore treating nurserymen in a miserly fashion after receiving plants from them. More interesting was the revealed assertiveness of the long-entrenched colonial scientific fraternity, some of whom were ready and able to use their status and scientific knowledge against the appointed professional to win an academic point, to indicate the superiority of their local knowledge or to express some personal animosity. A very small part of the enquiry proceeded thus:

George Macleay to Moore: 'Do you not also admit that a body of gentlemen might be found in Sydney with greater scientific attainments—I do not say it invidiously—but with a greater amount of knowledge than yourself, who might assist in naming the plants which already exist in the Garden?'

Moore: 'It is possible there may be; but I do not think you will find any body of gentlemen in the Colony, who are acquainted with the management of a Botanic Garden ...'

Donaldson revived the matter when questioning the Reverend George Turner: 'Should not the Committee consist of scientific gentlemen?' Turner shrewdly replied, 'Yes; of scientific, practical, and impartial gentlemen ...'.[66]

The outcome of the gruelling business was that the Select Committee submitted a report to Sir William Denison pointing out that while 'in the general management and arrangement of the Gardens, Mr Moore appears to have acted with ability and industry ... much has been left undone ...'. The labelling of plants should have been more correct and complete, a catalogue was needed, the accounts were 'kept in a very unsatisfactory manner', the Annual Reports were deficient, and there had been a 'too extensive distribution of plants and seeds from the Gardens'. The lecture room and library had not been used to the best advantage and the lectures themselves appeared 'to have been of a desultory character'. It was therefore recommended that the Director should henceforth be the Curator, 'subject to the control of, and responsible to, three Commissioners'. Thus, the politico-scientific fraternity did not want to lose Moore, but rather to control him.

The Governor did not agree, preferring the current arrangement whereby a responsible Director seemed clearly preferable 'to three unpaid and so far irresponsible Commissioners'.[67] That ended the matter, and Charles Moore thus emerged officially, if not professionally and personally, unscathed from his trial to embark upon a further forty years of service.

In January 1856 Moore received a copy of the committee's report together with some pertinent requests from the Governor—the completion of 'the catalogue of the contents of the garden ... with all necessary expedition'; the submission with the Annual Report of a detailed account of expenditure; the keeping of an accurate account of all plant exchanges; the presentation of a catalogue of the Garden library, with indications of how the books might be rendered 'useful for Public reference without risk of loss or damage' and a statement of the course of lectures which the Director proposed to deliver.[68] Thus Denison made it clear that although he had saved Moore from the control of another committee, he had taken careful note of the charges made, and accountability would be expected. In April Moore submitted for the Governor's approval the library list and 'Syllabus of the Lectures' proposed for the Spring.

By June 1856 the public had 'free access three days a week' to the library, to which funds were allocated. In 1857 the long-awaited census of plants in the Gardens appeared,[69] with a brief, unsigned introduction. During the enquiry it was maintained that instead of labelling each plant, it would be better to number the plants according to a published list of corresponding numbers and classifications. Fortunately Moore persisted with his direct method instead of adopting this clumsy procedure. The catalogue was arranged in alphabetical order of genera, with the species unnumbered. It indicated that, by 1857, the Gardens contained nearly 3000 species of flowering plants and ferns, comprising about 740 species indigenous to New South Wales, 110 species from elsewhere in Australia, 1860 species from overseas and 230 horticultural hybrids.

Notwithstanding the exhausting demands of the enquiry, other routine, unforeseen and even embarrassing matters still required attention. Moore was annoyed by a ship's hawsers being 'attached ... to the Trees growing in the Domain', and the Governor ordered 'measures for remedying the evil'; there was the problem of 'drainage from the Mint ... being discharged into the outer Domain', followed by the assurance that this was 'merely ... rain water'. Ironically, just after the enquiry, Moore was asked to supply information for the benefit of the South Australian Government, 'relative to the management of the Botanic Garden, Sydney' and there was even the request that he should arrange for the accusatory Mr Guilfoyle to receive two cases of Norfolk Island pines, ensuring that they received 'the same care ... as if they were for the Government'! A much happier commission was to supply seeds and plants for 'the Public Gardens and Squares' of Melbourne. Similarly, the advice that the Governor had approved the removal of stones 'from the old Limekiln, near Fort Macquarie ... for building the Boundary Wall of the Lower Garden' must have been welcome.[70]

Other matters for concern included the Government's approval 'of permission being given to the City Commissioners to remove gravel from the portion of the Inner Domain between the Government Stables and the Road leading to the Botanic Gardens for the maintenance of the Streets of Sydney'. Moore was to inspect this operation periodically, and within seven months he explained why, on his own initiative, he had prevented the work from proceeding.[71] There was also the alarming suggestion that the Sydney Fishing Company be allowed 'to land and sell Fish on the flat

Ground southward of the Jetty at Biggs Baths in the Domain', and perhaps to excavate a 'receptacle' in the sandstone in which to keep the fish alive. Predictably, Moore quashed this proposal with speed and resolution.[72]

Not surprisingly, Moore made no reference to the enquiry in his eighth Report. In 1855 he chose to emphasize the work done to improve the habitat of the 'Azaleas and Rhododendrons, a class of plants of considerable interest and beauty', using soil brought from the neighbourhood of Rose Bay. The perennial retaining wall had been extended, and silt, chiefly deposited by the Tank Stream at the head of Sydney Cove, had been used for filling part of the reclaimed area. Once again, the Colonial Secretary, Edward Deas Thomson, merited mention for his interest, this time in presenting a large and valuable collection of plants he selected in Europe.[73]

In the following year, as a result of the institution of responsible government, Moore reported for the first time to the Secretary of the Land and Public Works Department. He referred as usual to 'trenching and manuring' activities, and of course, to 'the sea wall', but emphasized a policy of 'thinning' which had caused the Gardens to assume a 'rich and varied aspect' not previously seen during his term of office. This operation,

although it caused at first a slightly naked appearance in some parts of the Gardens, has been of the greatest possible advantage, not only from the benefit which the remaining trees and shrubs receive by the circulation of more air, and the increase of light, of which they give ample proof by their extraordinary growth, which has entirely taken away the temporary bareness, but it has also been the means of opening out to view many beautiful native and exotic trees, which were before partly hidden ... To follow out this system generally will, in my opinion, be fulfilling one great object of a Public Garden. It is not by the acquisition of a vast number of species, or the crowding together of an endless variety of plants, possessing neither beauty or value, that the public taste is improved, or the cause of education served, but rather by a judicious selection and cultivation of such types of genera as will, while properly illustrating natural families, at the same time be interesting and instructive. Plants, therefore, remarkable for beauty, singularity, or their utility to man, are the most suitable for this purpose, and such it has been my object to select for, and retain in this Establishment.[74]

Having made one of his most significant policy statements, Moore noted that Mr Thomson's sugar-cane had flowered, some olive oil had been produced, experimental sowings of a *Festuca* grass had been made, and 'punts from the steam dredge' near Circular Quay, were being 'most advantageously engaged' in conveying silt to form a 'ground of an excellent description' in the Lower Garden.

During 1857 Moore continued 'to clear away the useless trees, whether indigenous or exotic, which have been allowed hitherto to encumber the ground ...' for he felt that the Garden was 'still crowded in many parts with uninteresting plants'. Clearing of a much different kind was necessary in the southern or upper part of the Garden, where the small sedge known as 'Nut Grass', *Cyperus rotundus*, had proved itself to be 'the greatest pest with which any garden can be infested'.

The work-force in the Gardens at that time comprised nine men working a six-day week for 6/6 to 7/6 per day.[75] They must have been kept vigorously employed as they carried out the Director's orders. One of the labourers was Anthelme Thozet (1826–78), the French-born naturalist who became a prominent botanical collector in Queensland.

Sir William Denison, the most 'Scientific Governor' since Brisbane, not only supported Moore after the enquiry and presented specimens for the 'Public Herbarium, but also 'most kindly authorized a portion of the Kitchen Garden attached to Government House to be fenced off [as] a secure enclosure ... for a nursery and propagating ground'. The Governor sanctioned a trip to the Blue Mountains in the summer of 1857–58, and Moore worked his way from Hartley back towards Lapstone Hill before crossing 'the line of range' to Kurrajong then following Bell's Line to Mount Tomah. The herbarium was thereby 'greatly enriched' and exchange material obtained.[76]

Moore's next important journey was to the Richmond and Clarence Rivers in 1861 to collect timber specimens for the London Exhibition of 1862, when he exhibited 115 samples with notes on classification, uses, locality, Aboriginal and vernacular names. It was to be claimed later that 'Woods Indigenous to the Northern Districts of the Colony'[77] (duly revised for subsequent exhibitions) was Moore's chief published work for forty-five years.[78] His friend (now Sir) William Macarthur, made similar collections from the southern districts for the Paris Exhibition of 1855 and the London Exhibition of 1862, producing catalogues of 240 and 193 species respectively.

Meanwhile, early in 1861 there was another parliamentary enquiry, this time into 'the present state and management of the Sydney Domain; more particularly as to the recent fencing off of a portion of the ground by a permanent fence'.[79] The problem arose when the Secretary of Lands, John Robertson, who had charge and control of the Domain and the Botanic Gardens approved an application from a 'Committee of gentlemen' seeking to fence a portion of the Domain set aside for cricket matches. The cricketers erected the fence at their own expense around the field a little to east of the present site of the State Library. By all accounts it was less than elegant, and one finding of the Select Committee, chaired by the Member for Wellington, Sylvanus B. Daniel, was that it ought to be removed, for it interfered with the right of the public to the free use of the Domain and was both 'unsightly and unnecessary'.

The ten-man committee examined sixteen witnesses, including, of course, the hapless Director, who although nominally a member of the 'Committee of gentlemen' responsible for the oversight of the use of the Domain for cricket, had not attended a meeting for eighteen months, and felt that the fence should be removed. Like other enquiries, it soon embraced an unexpectedly wide range of issues, many apparently quite remote from the terms of reference.

Much interesting information was revealed about the Domain, some being supplied by Moore himself as he answered in a direct, even breezy manner, batteries of questions on an astonishingly diverse range of matters. For example, it was stated that in March 1861 nine horses and thirty cows were authorized to graze in the Outer Domain. Moore's comments included:

'If no cattle were admitted it would take more than a thousand pounds a year to keep the grass down'

—and there would be the danger of fire.

'I am informed, almost all of the dairymen's cows in Woolloomooloo get into the Domain during the night and are taken out before the bailiff is up in the morning.'

'Oh, by-the-by there is a regulation that all dogs may be shot found at large in the Gardens, and another that grooms are prohibited from exercising horses in the Domain.'

Moore advised that six men were constantly employed in the Domain, including the Inner Domain and Government House grounds. He confessed that (after thirteen years in Sydney) he was not sure of the extent of the Outer Domain, although he had asked the Surveyor General's Office several times.

Further discussion concerned Hyde Park, the fence screening the City Corporation Baths, and the landing of cargoes from vessels in Woolloomooloo Bay. The Chairman suggested, 'They may be smuggled goods for all you know?' Moore replied, 'Yes, for aught I know'—or indeed, cared, it seemed.

The Director advised: 'The native trees are dying off fast, and unless we make some preparation there will be no trees in the Domain. I am in hopes, however, now, that before the gum trees are dead, the other young trees of a permanent nature will have grown up.' There was mention of the value of trees in 'skirmishing exercises' and of the obstacle that the offending fence presented to 'the evolutions of the troops or the volunteers'.

Then Victorian morality became an issue. Alexander Dick, Member for Liverpool Plains observed: 'Owing to the trees being destroyed it is quite possible for ladies driving past in their carriages to see naked bathers ...' Moore replied that there was a fence. He was asked whether the public were admitted to the Domain at night. 'They are,' he replied. Was this desirable? 'Well, I think it gives rise to a great deal of immorality ...' Richard Driver, Member for West Macquarie, pursued the issue:

Driver: Are there any immoral practices carried on in the Botanic Gardens after dark?
Moore: Not that I am aware of. The gates are shut.
Driver: But during the day have any complaints reached your ears?
Moore: Yes; some complaints have been made, but in most cases the parties have been turned out. They have never been absolutely found in the act, but where there has been a suspicion of it they have been turned out.

The Inspector General of Police, John McLerie, was apparently still in a state of shock over the matter. Recently, one of his inspectors had reported that 'the scenes of immorality he saw ... were beyond comprehension'. Gas lamps should be installed in the Domain, where he said, 'homeless female outcasts' lived 'about the drains and holes', adding with a regrettable choice of words, that they were disseminating 'more disease among the youth of Sydney than it is possible to conceive'.

David Wilson, overseer of the Domain and Hyde Park, agreed that lamps 'would be very beneficial', for 'a great many bad characters go and sleep all night in the bushes, and under the rocks, towards Lady Macquarie's Chair ... I see them go in at dusk, and out in the morning,' some of them comprising 'a parcel of little boys and girls, from ten to fifteen'.

George Kemp, Domain bailiff and sworn constable, believed that there were only 'half a dozen that I may term habitual idlers' who frequented the Domain. Generally those 'who lie about the Domain are decrepit old shepherds who come down ... from the country', spend their wages, and then loiter in the Domain or go to the Benevolent Asylum. He did not think there was much immorality, but visitors were told, when necessary, that they may not lie down; they must sit on the seats. In any case, gas lamps and a patrol were clearly needed.

Mr Kemp was something of a mathematician. When ultimately, the matter of cricket was discussed, he thought about 300 cricketers used the Domain. Being on his feet for fourteen hours a day, he had ample time to make observations. He noted that at times one person per minute passed through one of the entrances. There were five entrances, so the calculation of the annual number of visitors was relatively simple: 14 hours × 60 minutes × 6 days a week × 5 entrances = 25 200 people per week, or 1 310 400 in a year!

William Tunks, government trustee on the Domain Committee and well-known cricketer, sternly defended the social benefits of the game:

... once we have begun to play orderly cricket we have much better conduct, morally and socially, among our young men. We used to have a much larger number smoking and dissipating about the Domain, and a much larger number of children as a nuisance in the streets, flying kites, playing with peg tops, wearing out their clothes playing marbles, and so on. Now we get them to that kind of recreation which tends to develop their physical powers ...

Henry Parkes, then Member for East Sydney, was not convinced and in typical style declared:

The Domain is set apart for the recreation and the promotion of the health of the whole population, more especially invalids, young children of tender years, and sickly women, none of whom can be supposed to have any particular interest in the game of cricket; and if it were interfered with by the cricketers, however manly and thoroughly English their game may be, and however desirable it may be to encourage it, might not others claim also to have portions of it set apart ...

For example, local zoologists might like a section fenced off as a menagerie![80]

During the remainder of the decade, Charles Moore, his Gardens and his Domains, enjoyed increasing public and professional acclaim, generally untrammelled by charges or enquires. In 1860 the prominent Sydney physician-naturalist and first secretary of the combined Committee of Management, Dr George Bennett, who had once regretted that the so-called 'botanic garden' was 'merely a government vegetable and fruit garden', proclaimed that 'the Botanic Garden at Sydney is an object of great attraction'. The Garden had been 'lately materially improved by the present able Director, Mr Charles Moore'.[81]

The *Sydney Mail* of 28 July 1860 agreed:

The improvements that are being made in the Sydney Botanic Gardens are so gradual in their operation that they probably do not attract from casual visitors that attention which they deserve. Those, however, who remember the size and condition of the gardens a few years back will recognise the desirable alteration that has taken place ... perhaps the most noticeable is that in the lower garden, where a large quantity of additional land has been reclaimed, part from the bush and part from Farm Cove.

Three acres had recently been added to the Gardens, and three more would soon be added as the result of continued reclamation. The scientific section of this new ground would be laid out according to Dr John Lindley's system, with Exogenous plants (having 'leaves veined like network') represented by the four sub-classes— Hypogynous, Perigynous, Epigynous and Diclinous.[82] It was suggested that:

although the scientific division may not be properly appreciated by the public it will be of the greatest value to those who may wish to study botany. The larger number who last season attended Mr Moore's lectures at the Botanic Gardens than during

Rustic summer houses with roofs of Xanthorrhoea *leaves were built as early as the 1860s.*
(Mitchell Library, Sydney)

any previous season, may be regarded as an indication that the taste for this interesting science is on the increase.

For those visitors who enjoyed 'the beauty of the trelised [*sic*] enclosure along the eastern boundary of the upper garden walk' where one was 'encircled by roses, woodbines, passion flowers, and other popular varieties of climbing plants' for about a hundred yards, there was to be a new delight. Motivated by the offer of Alfred Denison, the Governor's brother, to present 'to the public his collection of birds', the Government had allocated funds for the aviary then being built in the midst of the trellised walk. Thus Mr Moore's responsibilities were about to be extended beyond 'the well-arranged plants and flowers' to 'the feathered aboriginal tribes of Australia'. His reaction was not recorded. When opened in Spring 1860, the aviary presented 'elaborate decoration' along its ridge-capping and a colour scheme of green and white to 'produce an extremely cheerful and pleasing effect'. The nearby walk had been

C. Cartwright's Map of the Governor's Demesne Land, 1816. It clearly shows
Mrs Macquarie's Road as well as the Gardens and plant nursery.
(Mitchell Library, Sydney)

The old: *a sketch of the Palm House, about 1890.*
(Mitchell Library, Sydney)

And the new: *the Pyramid Glasshouse, constructed in 1970–71.*
(L.A. Gilbert)

The Anderson and Robert Brown Buildings, 1984.
(David Bedford)

Residence built in 1874 for Charles Moore
(Director 1848–96). It was converted to offices in
1969 and is now known as the Cunningham Building.
(David Bedford)

Map of the Botanic Gardens, 1921.
(Royal Botanic Gardens, Sydney)

Botanic Gardens Creek, 1984.
(David Bedford)

The restored Gardens Restaurant in 1984.
(David Bedford)

The Lower Garden Pond, 1984.
(L.A. Gilbert)

Mount Tomah Garden in the Spring.
(Beulah Grewcoe)

View of the Gardens from the Premier's suite in the State Office Block, 1985.
(David Bedford)

'considerably improved by four composition figures, consisting of two lions and two Egyptian sphynxes' presented 'by Mr M'Quade of Potts Point'.

The initial bird collection included cockatoos, parrots, pigeons, 'sparrows and finches, fifteen varieties', Chinese pheasants, ducks and quails, English skylarks, blackbirds and thrushes, and someone bravely lent a 'black macaw'. Further donations were quickly made, and by Spring 1862 there were accommodation problems, which were partly resolved by the additional provision of 'two small octagonal houses ... neatly thatched with rushes'. The native birds now included Cape Barron geese, a pelican, a brush turkey and an albatross, and had been joined by a Chinese deer, an 'anteater' (presumably a South American species) and an unfortunate platypus, soon pecked to death by a brolga or native companion, which belied its name. Murray River cod had been introduced to one of the ponds, and it was sanguinely believed that as they had 'not been seen since being deposited', they were 'succeeding very well'.

Another two acres of reclaimed land had been added since 1860. Most of this new area was made into a lawn where it was proposed to grow the most ornamental of native forest trees. The systematic plots were being further developed, and for the convenience of botanical students there was now 'a rustic house ... octagonal in shape' thatched with Grass-tree leaves (*Xanthorrhoea*) and presenting 'a very sightly appearance'.[83]

In August 1863 Moore proudly advised Sir William Macarthur that he was 'glad to say that the Garden was never held in higher estimation than at the present' and added that he proposed 'to be absent from the Garden a good deal' during the summer of 1863–64, in order to collect 'in localities which I have not before visited. I am the more anxious to do this at once, that some of the fruits of my labors may be made available to Mr Bentham in his work on Australian Botany'.[84] George Bentham acknowledged Moore's collections more than 360 times in *Flora Australiensis*, 1863–78.

Although the press generally was becoming more frequently enthusiastic in its praise of the Gardens, Moore must have gained greater comfort when the *Horticultural Magazine* added its commendation in 1864:

It is a conceded fact that a nation's progress and the love of gardening, go hand in hand, and we should imagine that were a stranger to land in the Botanic Gardens, on his first arriving in this Colony, his impression would be that this country is in a progressive state.

The contribution to this pleasing situation by Moore and his former viceregal champion was duly acknowledged:

It must be patent to the minds of the generality of our readers, the great improvements that have taken place in the Botanic Gardens during the last ten years, in fact, we date its progress from the time of Sir William Denison's advent; and it must be a convincing proof of the great utility an energetic Governor is to a young country ... Too much praise cannot be given to the worthy Director and his subordinates, for the success that has attended their exertions to make the place what it should be, the chief garden of New South Wales.

Only nine years before, 'the worthy Director' had been charged as inept, unqualified and incompetent.

Despite the generous contributions of silt from Sydney Cove and Darling Harbour, the reclamation work dragged on, always it seemed, behind schedule. In June 1866 it was noted that:

The large area at the head of Farm Cove, which is only covered with shoal water at high tide, can scarcely be classed under the category of 'Botanic Gardens' ... The action of the sun on the refuse deposited in the mud generates a rank and sickening odour ... The fetid and loathsome slough is an eyesore ...[85]

On several occasions during his long term, Moore urged that the resources of the bush be carefully investigated. He wanted settlers, bushworkers and botanists to combine their knowledge for the economic advancement of the colony. In October 1864 he told the Philosophical Society of NSW:

The character of the vegetation of this Colony, in many respects so remarkable, is, as regards its economic value, but little understood. From it neither commerce, science, nor the arts have as yet been benefited to any appreciable extent. With the exception of a few trees, the timber of which is used for building and fencing purposes, scarcely any importance has been attached to any qualities of our indigenous plants, many of which I feel convinced contain valuable properties which only require to be made known.

Moore regretted that the botanical material sent to the Exhibitions of Paris and London had not stirred 'the authorities connected with these exhibitions' to investigate the properties of Australian bush products, although 'various vegetable substances, supposed to have medicinal, dyeing, and textile properties, were forwarded from this colony on both occasions'.[86] Clearly there was a need for a man whose science was wider than his botany, preferably one trained in chemistry and with some knowledge of analytical and experimental techniques. New South Wales, in fact, needed another Ferdinand Mueller or Joseph Bosisto.

———————————

Early in 1867 the faithful overseer and former superintendent, James Kidd, died in his cottage in the Gardens,[87] and further adversity followed. Moore had assembled 156 timber specimens from northern NSW for the 1867 Paris Exhibition. As one of the commissioners, Moore travelled to France, calling at Kandy on the way to see the garden at the Governor's residence. This diversion caused a late arrival in Paris, and he was relieved of his duties as a commissioner. Nevertheless he was warmly thanked 'for the great exertions in the arrangements and disposition of Articles' and extreme regret was expressed that the other commissioners should have been deprived 'of the further services of so valuable a Coadjutor'.[88] Moore found some compensation in visiting the south of France and Spain in the interests of the citrus industry.

In February 1869 Moore was back at Parliament House giving evidence before another Select Committee, this time deliberating on the Diseases in Grape-vine Prevention Bill which had been prepared to deal with a fungous disease attributed to *Oidium tuckeri*.[89] In May the Director, accompanied by William Carron and the celebrated orchidologist-surveyor, Robert David FitzGerald, joined the *Thetis* for Lord Howe Island, where an official enquiry was to be held. Many interesting discoveries were made, including the huge epacrid, *Dracophyllum fitzgeraldii* and the 'Kentia' palms, *Howea belmoreana*, *H. forsterana* and *Hedyscepe canterburyana*. Specimens were sent to Baron von Mueller who named an orchid in the collection *Dendrobium moorei*.

In the Spring of 1870, the Cook Centennial Year, it was considered that 'a good sized volume would be required' to describe 'everything interesting or pretty in the Botanic Gardens'. Improvements had been made in the walks, some of the lawns had been broken up with new beds of flowering plants, landscaping had been carried out on the slopes, and 'several hundred of loads of rock' had been carted out and replaced with 'town rubbish'. Citizens were urged to enjoy their leisure amid the fragrance being produced by roses, magnolias, wistarias 'and a hundred other plants'. The Domain was not similarly praised, for although Araucarias 'and other ornamental trees' adorned part of the ground, the western area appeared to have been altogether neglected with some of it 'little better than a bog' owing to water still streaming from the Mint. Although this provided 'probably the most frequented route of any between the city and Woolloomooloo', it was dangerous during the Summer 'on account of the cricket', and at other times it could not be used 'on account of the mud'. It was hoped that the Director would act to correct this 'perfectly disgraceful' state of affairs.[90] It was well that the Gardens were in good order to receive the influx of visitors to the city for the Intercolonial Exhibition held in Prince Alfred Park during September. Moore contributed a non-competitive display of living plants to the exhibition, and an informative chapter 'On the Woods of New South Wales' to the exhibition catalogue.

After a lapse of thirteen years, Moore favoured the government with 'a General Report' on the Queen's Birthday, 1871. Naturally there was much to record. The 'Upper Garden', (i.e. the present Middle Garden) 'the oldest and best protected' contained the greatest number of species, but the removal of unsightly or useless plants had continued. Fortunately, the two largest trees of the Norfolk Island Pine in the centre of this area had apparently not affected 'any of the plants growing near them, although they are the largest and oldest of the trees, and really magnificent specimens—doubtless the finest by far of any in cultivation'. It was estimated that one was 'now 120 feet high'. Unfortunately, while every care had been taken with the rhododendrons along the Macquarie wall, nearly all had perished, while the azaleas flowered 'every year most profusely'. In the western part of this Garden, rainforest trees such as Silky Oak, *Grevillea robusta* and 'the so-called "Moreton Bay Pine"' *Araucaria cunninghamii* 'collected and planted here principally by my predecessors, Cunningham and Fraser' had developed into fine examples. A Queensland Kauri Pine, *Agathis robusta*, planted in 1853, was now 53 feet high, while the palm plantation, begun in 1862, was flourishing. The palm groves of the Middle Garden continued to 'give a very pleasing effect', and indeed to many visitors, are still 'perhaps the most attractive feature'.

The Lower Garden had been laid out in a more or less irregular form with plantations and individual trees dotted over the extensive lawns. Much quarrying had been done and large quantities of fresh soil had been introduced to render this formerly 'barren, undulating, rocky' and 'rather unsightly' area into 'one of the most beautiful parts of the Garden'. The western section of the sea wall, near the Governor's bathing-house, had been raised by about 4 feet for 130 yards, to protect the vegetation from the 'most blighting effect' of the spray generated by strong north-easterly winds, and of course, the reclamation work continued.

The five acres of the old kitchen garden of Government House had been transformed

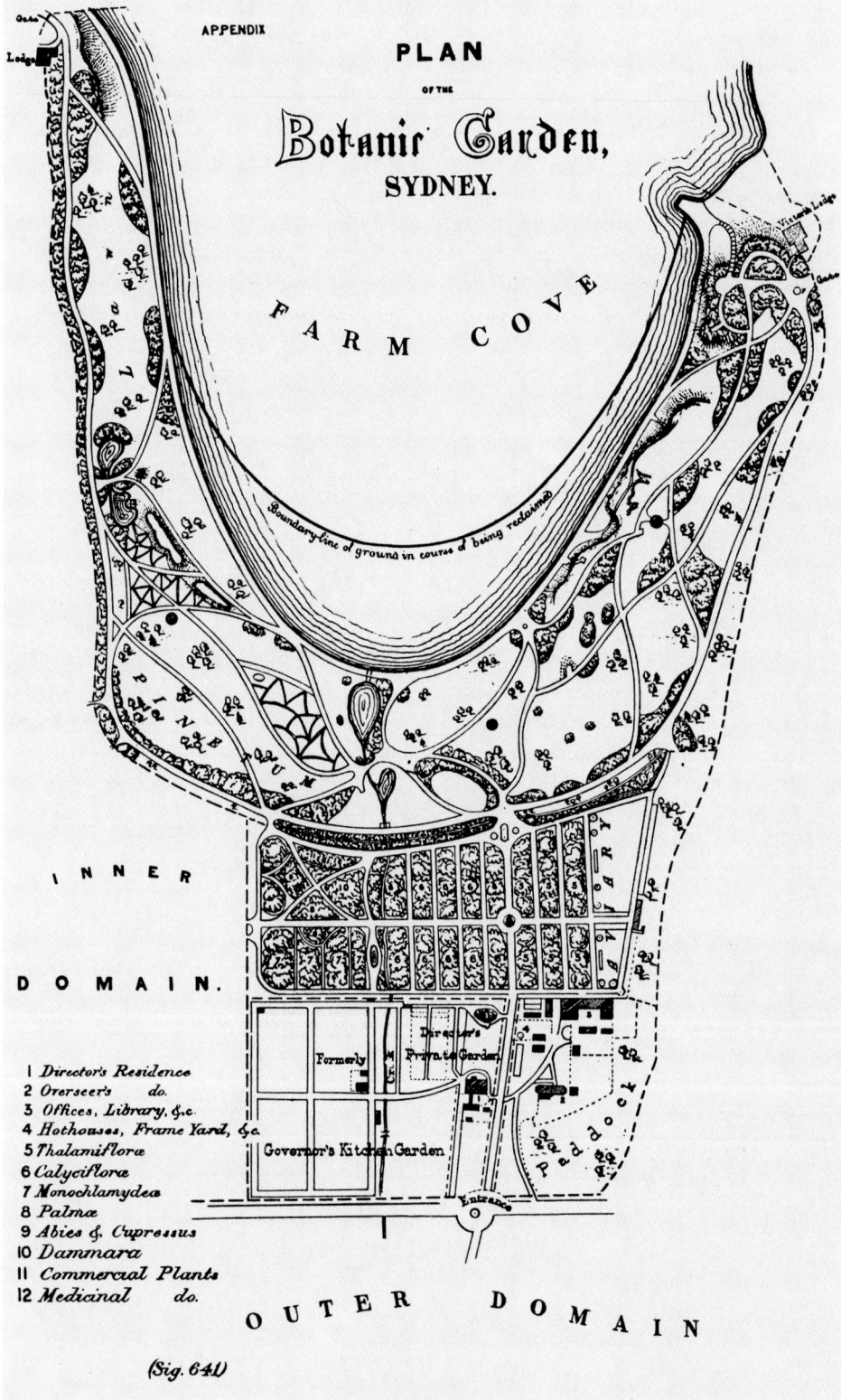

APPENDIX

PLAN

OF THE

Botanic Garden,

SYDNEY.

FARM COVE

Boundary-line of ground in course of being reclaimed

INNER

DOMAIN.

1 *Director's Residence*
2 *Overseer's   do.*
3 *Offices, Library, &c.*
4 *Hothouses, Frame Yard, &c.*
5 *Thalamifloræ*
6 *Calycifloræ*
7 *Monochlamydeæ*
8 *Palmæ*
9 *Abies & Cupressus*
10 *Dammara*
11 *Commercial Plants*
12 *Medicinal   do.*

Formerly

Director's Private Garden.

Governor's Kitchen Garden

Entrance

Paddock

OUTER   DOMAIN

(Sig. 641)

into a nursery and a source of 'greenstuff for the animals and birds now in the Garden', but with additional expenditure the area could be made 'a more integral part of the Botanic Gardens', and in fact it became a substantial part of the area now known as the Upper Garden. Continued attention was being paid to medicinal and commercial plants, especially to a rather extensive collection of sugar-cane varieties.

After more than a decade, the aviary remained a source of attraction to visitors. However, 'it has ... been found to be expensive and difficult to keep up the interest of the collection, in consequence of the losses sustained by deaths and the mysterious disappearance of some of our rarest native birds'. Although donors to the aviary were 'entitled to such plants as can be spared', their number had recently declined. Nevertheless, the zoological collection had become more diversified to include a kangaroo, a wallaby, two bandicoots, a white kangaroo rat, two koalas, a seal, a possum and a mongoose. The birds now included an emu, a wedge-tailed eagle, a brush turkey and a New Zealand penguin.

A plant exchange programme had been maintained, and valuable additions made to the library and herbarium, but the condition of nearly all the buildings was a matter for concern:

The office or library, the quarters of the Bailiff and Propagator, adjoining the storerooms, the Overseer's house, and even that in which I reside, are all unsightly. With the exception of the latter, all are old, badly built, in wretched condition, damp, most unhealthy, and quite unworthy of the place. To level every one of these with the ground and replace them by others more suitable, in better positions, and lay out the ground which they now occupy in an ornamental manner, are works which urgently require to be done, and which doubtless the public would view with the utmost satisfaction.

In the Domain, there were now 'nearly four miles of carriage roads, and upwards of four miles and a half of footpaths'. Maintenance of these was expensive, having now to be carried out with blue metal, whereas formerly gravel from the Outer Domain was used.

Regrettably

the native trees, mostly Eucalypts and Banksias, which a few years ago grew so thickly in all parts of the Domains, are fast disappearing from natural decay. However much the loss of these trees is to be deplored, there was no means of saving them. To replace these by others of more permanent growth has been an object steadily held in view for some years past.[91]

Thus there appeared increasing evidence of what Australian Museum curator Robert Etheridge later described as 'Mr Moore's predilection for that scourge of gardens, the Moreton Bay Fig ...'.[92]

It is interesting to consider some of the people associated with these developments in the Garden. For example: John Connor, who in February 1871 carted '44 Loads of Manure' to the Gardens, and another 64 loads in April; John Neville, who during the first three months of 1871 supplied '90 pints of milk for Aviary' to supplement considerable amounts of canary seed, biscuits, wheat, maize and potatoes; John Grant,

---

Opposite: Plan which accompanied Charles Moore's Report of 1871. Elsewhere he noted that by June 1872 the 'Area of Reclaimed Ground Lower Garden Between Old & New Sea Wall = 7 acres 251 sq. feet.'

who levelled 526 cubic yards of filling in the reclaimed area at 9d per yard, thereby earning £19-14-6 in July 1871; Henry Bell who in July supplied 'Birds' meat' and H.E. Haffield, who in September, painted, grained or varnished 116 garden seats at 3/6 each.[93] In mid-1871 Thomas Sutcliffe Mort presented a glasshouse, in the form of an 'irregular octagon'. Fitted up by William Huntley, who worked for $26\frac{1}{2}$ days at 8 shillings a day, this new building was 'a conspicuous ornament near the principal entrance' and by the Spring, was 'daily accessible to the public'. The donor also provided plants for the initial display, including ferns, begonia, azalea, rhododendron, hibiscus, anthurium, dracaena and coleus.[94]

The 'principal entrance' itself was soon rendered more welcoming and impressive by the installation of a fine set of wrought iron gates from Robert Dunlop's workshop in Forbes Street, Woolloomooloo (costing £145) supported by massive, but finely-carved sandstone piers from Messrs Hanson & Sharp. The ironwork was painted mauve, with the crown and Royal monogram in gold. The *Illustrated Sydney News* of 28 August 1873 featured a woodcut of the new gates, and declared 'when the whole work is finished it will present a marked contrast to the dirty and dismal appearance of the old gates, which disgraced, until recently, the principal entrance to our beautiful Botanic Gardens'. In three years the gates assumed a new significance when the area now known as the Upper Garden was generally opened to the public for the first time. Previously it had been a more or less closed section containing the old Kitchen Garden, propagating beds, stables, residences and other service facilities. In 1962, when construction of the Cahill Expressway destroyed the magnificent Fig Tree Avenue between the Gardens and the Outer Domain, the gates were moved a short distance to their present position facing Mrs Macquarie's Road.

Granted six months leave on half-pay in 1874, Moore joined his brother David at the Botanical Congress and International Horticultural Exhibition at Florence. Five years later he favoured the Under Secretary for Lands with his last General Report. The next Report did not appear until after his retirement. Perhaps, in view of Moore's belief that the Gardens had 'now arrived at that state of completeness which, so far as the laying out is concerned' required 'little more to be done', he also felt that there was little more to be said. After thirty years of employing 'a vast amount of labour' on reclamation and on 'the parapet wall which has been erected round the bay', Moore had at last completed his grand project. Despite the salinity of the filling material, the Lower Garden lawns now presented an expanse of 'deep dark green' which had 'never before been so remarkable'. Where salt water could be excluded, ponds made in the reclaimed ground had proved successful for the growing of aquatics including various water lilies. There was now 'a large and valuable selection of medicinal plants and those yielding substances of commercial value' and elsewhere in the Lower Garden attention had been given to 'groups of special plants, such as agaves, aloes, cycads, cactuses, yuccas, &c.' thereby giving 'a botanical interest to the garden which it would not otherwise possess'. Many conifers and most palms were growing well, while the three great Norfolk Island pines were now estimated to be 94, 111 and 112 feet high. Hoping to induce 'colonists to grow plants of economic value and easy culture', Moore obtained seeds of jute, tobacco, castor oil plant, 'three very distinct kinds of millet', flax, safflower and indigo. Along Botanic Gardens Creek, six dams or weirs of cut stone had been constructed, to make the watercourse 'permanent and sightly' and to provide some storage. It was believed, rightly, that 'when surrounded by proper plantations' the creek would 'be an ornament to the garden', and so it is today.

The Director pointed to the need for plantations of timber trees and suggested the establishment of a government nursery to enable 'a commencement towards the re-foresting of the country'. In 1881 a State Nursery was established on 22 acres at Campbelltown, with Alexander Moore as first superintendent.

Following previous recommendations, 'the old Barracks, in which were the office, herbarium, library, and quarters for employees, as well as the stables attached', were pulled down in 1878, 'and new buildings of a substantial character erected instead'. The new buildings comprised

a museum, lecture room, library, office, storeroom, and two separate dwellings for men of the establishment. Stables for horses employed in the Gardens and Domains have been placed in the old kitchen garden, as much out of sight as possible and away from public resort. A new house has also recently been constructed on the eastern side of the garden, for the Overseer ... The building for ... a museum is sufficiently large for present purposes, and perhaps may suffice for some years ... but certainly it is not commodious enough to contain all the interesting exhibits of vegetable products which it is possible to obtain ...

Moore's friend, Dr George Bennett, had selected some terracotta vases 'at the establishment of Messrs. Doulton & Watts, Lambeth, London' and these would soon enhance the Gardens. Much had been done to ensure the health and comfort of the birds and animals, but there had never been the intention to 'form ... a general zoological collection'. Nevertheless, 'this adjunct to the garden is ... of sufficient interest to amuse and attract a very large proportion of the visitors'.

In the Domain there had been the problem of providing footpaths on which people could walk with comfort. As every plan short of asphalting and tarring had been tried and found wanting, some asphalting of paths had been professionally done at a shilling per yard, while tarring had been carried out by regular staff 'at much less cost'. The replanting programme was already providing excellent shade and an 'ornamental appearance' in 'this favourite public resort', more especially Moreton Bay Fig, *Ficus macrophylla*, and allied species; Brush Box, *Lophostemon confertus*; the Asian Camphor Laurel, *Cinnamomum camphora*, and English oaks, elms, poplars and plane trees.[95]

Early in 1878 it was decided that the first Australian International Exhibition would be held in Sydney under the auspices of the Agricultural Society of NSW.[96] This 'International Exhibition of Works of Industry and Art' quickly became a government responsibility, and the Colonial Architect, James Johnstone Barnet, was instructed in December 1878, 'to prepare plans for a building suitable for an International Exhibition, proposed to be built in the Inner Domain'.[97]

With remarkable speed, Barnet presented a design for an enormous structure, some 800 feet by 500 feet, with a ground area of $5\frac{1}{8}$ acres and a total floor area of over $8\frac{1}{3}$ acres. On 2 January 1879 Barnet marked out the site between Governor Bourke's statue and the Government House Stables on land which had been 'exclusively used by the Governor for grazing purposes, to which the public had then had no access whatever', and work began in eleven days.[98]

The Garden Palace as it appeared about 1880 from the pond in the Lower Garden.
(National Library of Australia, Canberra)

Such was the demand for space that in addition to Barnet's gargantuan building in the Inner Domain, a further twenty-three acres of the Outer Domain were appropriated, despite opposition, for machinery, livestock and other exhibits requiring considerable space. Moore was charged with the task of landscaping the grounds around the 'Garden Palace' in January 1879, and by the end of the month twenty-seven men were employed on the project at 7 shillings a day. By the Spring, the number had risen to over sixty.[99] When the Governor, Lord Loftus, officially opened the exhibition on 17 September, the 'immense labour' of draining, filling, turfing and planting the area with some 28 000 seedlings and shrubs in gardens and borders, had been completed.

The palace itself was an astonishing example of Victorian architectural exuberance, with towers and turrets deployed around a giant dome, 100 feet in diameter, surmounted by a lantern over 200 feet above ground level, forming a cavernous canopy over 'a bronze statue of Her Most Gracious Majesty Queen Victoria' set at the intersection of the nave and transept. The materials included 4 500 000 superficial feet of timber, 2 500 000 bricks, 243 tons of galvanized iron, and vast quantities of glass, statuary and ornamentation. Partly because of a pioneer use of electric light, this cathedral of art, science and industry was built in eight months, with work proceeding at night.

Notable in the vast array of exhibits were the collections of timbers and other natural history displays, such as the palaeontological specimens of the recently deceased Reverend W.B. Clarke. When the exhibition closed on 20 April 1880, over a million people had been admitted.[100] Thereafter, sections of the edifice were used for meetings and concerts, the storage of government records, the Art Society collections, the Linnean Society's headquarters, and a museum. Then, early on the morning of 22 September 1882, the entire Garden Palace and virtually all of its contents were destroyed in one of the most spectacular fires in Sydney's history. Among the losses were field records and other materials from the Reverend W.B. Clarke; the office of the newly created Forest Conservancy Branch, the Art Society's 300 paintings, the entire library of the Linnean Society and 'the valuable collection of plants belonging to the members'. Charles Moore lamented that twenty or thirty thousand of his ornamental plants around the palace were nearly all destroyed.[101] Between 1880 and 1882 work gangs of seven or eight, under the superintendence of such able men as Frederick Turner and Julius Henry Camfield[102] had kept the Garden Palace grounds in order while immense numbers of visitors tested both skill and patience. After the fire came the task of clearing the debris, amid which a slight young man named Maiden would have been seen earnestly rummaging.

From the practical and aesthetic viewpoints, the four-hour holocaust removed from the Sydney skyline a huge building which would have been difficult to utilize fully and to maintain effectively. From the historical and scientific viewpoints, the fire was a disaster. For Charles Moore, the loss of the palace meant the ultimate annexation of some nineteen acres to the Botanic Gardens.

In 1938, to mark the sesquicentenary of European settlement, a sunken garden was established on the site of the great dome as a Pioneers' Memorial, and a century after the exhibition was opened, the Governor of NSW, Sir Roden Cutler, unveiled a plaque comemmorating the Garden Palace and the exhibition it housed. An earlier memorial to the palace was completed in 1889 in the form of a fine gateway with beautifully carved Waverley sandstone piers supporting wrought-iron gates incorporating a representation of the palace dome. In 1962, when the Cahill Expressway caused such upheaval, the gates were moved from their original site in front of the State Library to Macquarie Street.

During and after the excitement of the International Exhibition, other work continued. The zoological collection was enhanced by the construction of a monkey house in 1880, just three years before most of the collection was transferred to the Zoological Society for stocking the new Zoological Gardens at Moore Park. However, the birds were held in such public regard that some were returned to the Gardens Aviary, which remained until about 1940.

In 1883 Moore's forty page *Catalogue of the Botanical Museum* was published, listing 763 exhibits displayed in bottles and cases—timbers, leaves, barks, seeds, fibres, fossils, plant extracts. Next year Moore published his *Census of the Plants of New South*

Wales, in which he relied 'principally' on Baron von Mueller's *Systematic Census of Australian Plants* of 1882, and followed the nomenclature used by Bentham in *Flora Australiensis*. The aim was simply 'to provide a ready means of reference to the systematic names and authorities of all plants found hitherto in a natural or naturalized state in this Colony'. In 1883 and 1884 Moore read papers to the Royal Society of NSW on the genera *Macrozamia* and *Doryanthes*, followed in 1890 by some brief 'Remarks on a New Plant rich in Tannin', an American species of dock, *Rumex hymenosepalus*.

Some of the Director's other activities during the early 1880s were not so creative. In January 1882 Moore laid charges against Captain Richard Ramsay Armstrong, RN, magistrate and administrator on Lord Howe Island. The point at issue was relatively trifling, involving the sum of £30 and a supply of some Lord Howe Island material, chiefly palms. Moore claimed he received too little for the sum given, while Armstrong claimed the consignment was worth much more. By April 1883 Moore was back before a Select Committee examining a variety of complaints against Armstrong.[103] Later in the year, Moore adopted a sternly moralistic attitude towards Mr and Mrs J.C. Dunlop, whom he considered were displaying 'uxorious affection' in the sanctity of the Gardens. Using language described as 'calculated to cause a breach of the peace', Moore had the couple ejected. Dunlop successfully sued Moore in the Water Police Court, and the Director appealed to the government. The Premier and Colonial Secretary, Alexander Stuart, took the most unusual step of reversing the magistrate's finding, and announced, 'Mr Moore has my entire sympathy', for he considered there had been 'an entire miscarriage of justice'. The case should have been dismissed, he said. Naturally there were misgivings in parliament, where the Member for Upper Hunter, John McElhone, insisted that 'Mr Moore, who was only a Civil servant, deserved all he got, and a great deal more'.[104]

It was timely that under the 1884 Crown Lands Act, revised and typically stringent regulations for the Gardens and Domain were promulgated in December 1885. First:

No person in a state of intoxication, or of reputed bad character, or who is not cleanly and decently dressed, shall enter or remain within these gardens; and no person shall behave in an improper or offensive manner, or use bad language, or commit any act of indecency therein.

Furthermore, children under twelve were to be accompanied by an adult; no races or games were to be played, nor 'any stone or other missile' was to be deposited or thrown; no fires were to be lit; no garbage was to be left; smoking was not permitted, and flowers were not to be carried into the Garden. Dogs, goats and poultry were all forbidden entry; articles were not to be sold; walking on grass which bordered paths was not permitted. Garden furniture and ornaments were not to be defaced or marked, and plants and animals were not to be taken, injured or disturbed. Seats and fences were not to be climbed or hurdled, and both standing and lying on seats were forbidden, as was lying 'on the grass near to any of the walks'. The permission of the Colonial Secretary was required for any meeting of twenty or more, and there were provisions for expulsions and fines. The hours of opening were 6.30 a.m. to 7 p.m. in Summer and 7 a.m. to 5 p.m. in Winter, and people found in the Gardens at other times would 'be forthwith removed'. Similar caveats were issued for the Domain, with

The popularity of the Gardens is shown in this photograph by John Paine, taken about 1880.
(Mitchell Library, Sydney)

special provision for the use of the roads by riders and drivers and for public meetings, which, however were not to be addressed 'in violent and unseemly language, calculated to inflame the minds of the hearers or cause a breach of the peace'.[105]

There was, of course, an immediate reaction. Why should not a person recline on a seat if this did not annoy others? asked one irate citizen. After all, 'invalids sometimes find the position essential'. Why should not people train for a race or a game? 'It is done every day.' Why should not people carry flowers into the Garden? 'Most ladies do so.' Why should not people lie on the grass near the walks? 'Picnic parties do it every day.' Who was the Director to say whether violent or inflammatory language was being used? '... is the director, who presumably is a good gardener, but not necessarily a good censor or critic, the proper person to exercise an opinion?'[106] There must have been many times when Charles Moore regretted holding what is glibly termed 'public office'.

At the end of December 1884 James Jones, overseer of the Domains, recorded a 'Diamond Drill at work in Inner Domain in connection with proposed underground railway'. The Director, who like many other citizens of Sydney, did not live to see this

*The water fountain near the Palm Grove was well patronized on a warm afternoon in the 1880s
when John Paine took this photograph.*
(Mitchell Library, Sydney)

long-term project realized, left on 15 January for New Zealand 'on a 6 weeks holiday
principally as an antidote for rheumatism & sciatica'. He resumed duty in March,
shortly after the departure of the Australian contingent for the Sudan.[107]

Moore's professional career was nearly at its end when in 1893 he published with
the able and crucial assistance of Ernst Betche, his most significant work, *Handbook of
the Flora of New South Wales*. Prudently, the advice of Baron von Mueller was sought.
His assurance was characteristic—warm, but firm: 'I will be happy to aid in the
elaboration of the Flora of New South Wales, but in order that no clashings or con-
tradictions occur in naming, characteristics and systematic disposition, it would be
necessary, that my *Census* should be the basis of operation.' Predictably, Mueller
considered it:

not at all necessary, to follow the arrangements and nomenclature adopted for the
Flora Australiensis, because hardly any one, who will use the special work on New
South Wales, will use the seven volumes of all Australia.
...
Furthermore if all the naming of orders, genera and species had to be strictly adopted
from the Flora Australiensis, the rule of priority, which finally must prevail, would be

carried out very imperfectly; and if the limitation of genera & species had also to be exactly in accordance with the Flora, all the research of the last 25 years (not only by me but also of European Botanists) would be lost sight of, and the work be so much *behind the times*!

Accordingly, 'Mr Betche, under your direction' might send to Melbourne 'the manuscript of order after order in the sequence of the Census' and Mueller would 'add … localities and *perfect* the whole in some other ways' before it went to press. Mueller also strongly advised Moore not to go to the printers before the beginning of 1888

so that efforts may be made, in which Mr Maiden wishes to share by help of his Department, to get the plants of the *remotest N.W. of N.S. Wales* next spring. I am satisfied that many genera and a very large number of species … from thence are yet to be added to the Flora of N.S.W.

If this were not done, 'the new Flora of NSW will at once be very incomplete'. Such safeguards could 'only be an advantage to yourself',[108] he warned.

Moore used Mueller's systematic arrangement, for it seemed 'to approach more nearly the ideal of the natural system than De Candolle's or Jussieu's system used in most colonial floras' and the handsomely bound volume, with its historical introduction, list of authors' names and botanical glossary, remained the chief work of its kind for over sixty-five years.

The influence of the Gardens and their Director became increasingly wide during Moore's time. Originally appointed to direct the affairs of the Botanic Gardens and the Domains, Moore also became associated with Hyde, Victoria and Wentworth Parks, a succession of intercolonial and international exhibitions, the grounds of Government House, Admiralty House, Garden Island, Sydney University, various gaols and court-houses and the gardens of stations being established along the rapidly-expanding railway system. Moore was appointed one of the original trustees of National Park in April 1879, and assumed four additional major responsibilities when he was well over sixty—the establishment of the State Nursery at Campbelltown (1881), the incorporation of the Garden Palace Grounds (1882), the modification of part of the Outer Domain to accommodate the 'New Art Gallery' (1885) and the inauguration of Centennial Park (1887). This work was applauded when Moore retired, when it was noted that he was also instrumental in introducing 'the Jackaranda, the pepper tree, and the gorgeous Hibiscus', which no doubt were well represented in the 35 000–40 000 trees and shrubs he had distributed 'to various public establishments throughout the colony' during his last decade in office.[109]

In meeting these responsibilities, Moore had the necessary support of a succession of competent men. Some, like Anthelme Thozet, William Carron and the loyal James Kidd, one-time acting superintendent, sent to the Gardens as a convict labourer in 1830 and superannuated as overseer in 1866, have already been noted. Others included John Duff (b. 1845) who succeeded Kidd as overseer on 1 September 1866 and became Inspector of Forests in December 1882; George Harwood (1842–1915) an accomplished nurseryman and landscape gardener, who joined the Gardens staff in September 1873, became overseer in January 1885 ('superintendent' from November 1891) and remained so until he died on retirement leave; James Cook (b. 1835) who joined the Domain staff in 1867 at 7 shillings a day and remained until 1907; similarly Patrick Harkins (b. 1837) from 1870 to 1903; John McEwen (1842–1913), super-

Parts of the Middle Garden still offer the same leafy shade that is shown in this photograph,
taken by Henry King in 1895.
(Royal Botanic Gardens Library)

intendent of the Campbelltown Nursery, 1884–1912; Ernst Betche (1851–1913) botanical collector 1881, botanical assistant 1896, chief botanical assistant 1908 until his death in 1913; Charles Peters (b. 1838) bailiff of the Gardens, 1866–96, 'ranger' until 1908, thereafter temporarily employed as 'Watchman on Moonlight nights' at 1 shilling an hour and on other duties until 1917. There were, of course, scores of other faithful workers during Moore's extraordinarily long term, many with careers which extended well into J.H. Maiden's time.[110]

By May 1896 Charles Moore's retirement had been 'decided by the Public Service Board', and the Director relinquished 'most reluctantly the charge ... he ... had for so long of the Botanic Gardens, the Centennial Park, and other public reserves'.[111] He had, in fact, stayed rather too long. Now a widower who had just turned seventy-six, Moore visited England and Ireland, returning to the Trinity College Gardens in Dublin, where he recalled 'the men and plants, etc. he had met or seen there in 1832'.[112]

Returning to his home, 'Myola', 4 Queen Street, Woollahra, close to Centennial Park, Moore was cared for by a niece, Margaretta van Heuckelum, until his death on 30 April 1905, just short of his eighty-fifth birthday. He had arrived in a pre-goldrush colony, and had seen it grow into a State of a new Commonwealth; he had, it was

112

The Gardens Tea Rooms, about 1890. After a fire in 1977 this section of the building was incorporated into the new Gardens Restaurant.
(NSW State Archives)

said, seen his beloved Gardens become 'the home of a collection of trees, shrubs, plants, and flowers ... 20 times as large' as that in existence on his arrival.[113] Tall, dignified Charles Moore, with his charming and 'commanding but genial manner,'[114] had long been a Fellow of the Linnean Society of London and of the Royal Horticultural Society, and a member of the Royal Botanical Society. He had served the Royal Society of NSW and its predecessor since 1856 as member, councillor, vice-president and president. On behalf of that Society, Professor T.W. Edgeworth David declared that Moore's 'good nature and courtesy' were

as well known to the public as to his colleagues on the Council of this Society and on the Board of Trustees of the Australian Museum. While Mr Moore has merited the gratitude of the people of New South Wales for having added so much elegance and beauty to their capital city, he deserves the special thanks of this Society for his long and useful services ...[115]

On the day the former Director was buried at Rookwood near his old friend Dr George Bennett, his successor, J.H. Maiden, proclaimed that for 'over 48 years, Mr Moore occupied the post of Director of the Botanic Gardens, to the advantage of the country and with credit to himself'.[116] It was a restrained tribute, but true enough.

113

Chapter 6

THE MAIDEN ERA
1896–1924

… my references to a botanic garden are meant in the full sense of the word, and not in the maimed or restricted sense of those gardens which are mere parks or horticultural establishments. Sydney is a capital city and her botanic garden is truly a botanical establishment; it is also one of the oldest in the world.

JOSEPH HENRY MAIDEN, 1912[1]

In his Presidential Address to the Royal Society of NSW in May 1880, Charles Moore referred again to one of his favourite themes by 'drawing attention to the very great necessity which now and has long existed of ascertaining the uses and economic value of the Australian flora. The knowledge which we possess of the properties of the greater number of the plants of the Colony is most imperfect'.[2] Before the year ended, Moore's successor-to-be, J.H. Maiden, had left England for Australia, where within eight years he produced *The Useful Native Plants of Australia* as his first major work, a 700-page volume which went far to correct the situation which Moore so lamented.

Joseph Henry Maiden, eldest son of Henry Maiden, an accountant, and Mary, née Wells, was born at St John's Wood, London on 25 April 1859. At the City of London Middle Class School (and later at the Birkbeck Institution) he developed skills in the three Rs, learned about workshop practice in manual training classes, and showed a propensity for science, especially chemistry. Selected by a visiting teacher, Professor F. Barff of Cambridge, he served as demonstrator for three or four years, becoming principal laboratory assistant and curator of the school's museum of raw materials imported through the Port of London. Although offered a Cambridge scholarship by the Fishmongers' Company of London, Maiden declined it, apparently on health grounds, to remain at home and study science at the University of London at which he matriculated in June 1879. He met pioneers of the technical education movement when demonstrating for 'Science made easy' courses conducted for city working men, and was nominated for a post in the Royal Arsenal laboratory at Woolwich. It is

Part of the old museum (now the Visitor Centre) about 1890. Some of the specimens still exist.
(NSW State Archives)

unlikely that such potentially bellicose work would have suited Maiden's taste, but he suffered 'disappointment after disappointment through illness', and late in 1880 was persuaded to take the traditional sea voyage to a warmer climate.

Maiden arrived in Sydney with a return ticket, a letter of introduction to a bishop stationed 'far from Sydney', and another from Professor Barff to Professor Archibald Liversidge of Sydney University. Liversidge was a fellow-advocate of technical education and a foundation member of the recently-established three-man Committee of Management of the new 'Technological, Industrial and Sanitary Museum of N.S.W.'

The museum's collection comprised a huge and diverse residue of specimens from the Sydney International Exhibition of 1879, and was housed 'in over one acre of space' in the Garden Palace. Liversidge had the first curatorship offered to Maiden, who began the enormous task of sorting and arranging early in October 1881.

In characteristic fashion, Maiden worked assiduously and effectively, and the museum promised to be ready for inspection in December 1882. On 22 September, however, the Garden Palace and Maiden's immediate hopes were destroyed together in the disastrous fire. When the ashes cooled, the irrepressible Maiden salvaged what he could and prevailed upon the Colonial Secretary to permit him to convert part of the Agricultural Hall in the Domain for museum purposes. This 'long galvanized iron building, unlined, and floored with battens' had been constructed for the exhibition behind Sydney Hospital, which had commandeered the southern half for out-patients. The northern half had just been vacated by the unfortunate Italians of the notorious Marquis de Rays's expedition. While they made a new home on the Richmond River, the lively curator cleared out their 'bunks and some rough screens' to make a new home for his Technological Museum. When Maiden, still not thirty, completed his *Useful Native Plants* in January 1889, the museum contained 'over 25,000 specimens' and by June 1891, 30 000.

The young traveller of 1880 rapidly became a well-assimilated colonist. On 30 November 1883 at Holy Trinity Kew, Melbourne, he married Eliza Jane (Jeanie) Hammond, youngest daughter of the late John Hammond of Manchester, and by July 1893, was the proud and devoted father of a son, Harrie Hammond, and four daughters, soon dubbed 'the beautiful Maidens', Gertrude, Mary Emily Eileen, Acacia Dorothy and Nellie Russell.

Maiden joined the Linnean and Royal Societies of NSW in 1883, favoured both with the first of scores of papers in 1887 and ultimately served both as president for two terms. Still eager to pursue his science studies, he enrolled at Sydney University, attended zoology lectures and sat for examinations in March 1885, before a further breakdown followed by a bout of typhoid fever ended the venture. He developed his correspondence with Baron von Mueller, attended Charles Moore's botany lectures, became the pupil, friend and admirer of the pioneer teacher and botanist of Parramatta, the Reverend William Woolls (1814–93) from whom he learned much about Eucalypts, and enjoyed informative bush excursions with the doyen of the Linnean Society, Joseph James Fletcher (1850–1926).

In 1883 Maiden established a herbarium in his museum, storing material in 'the same class of box' he was using forty years later in the herbarium of the Botanic Gardens. Maiden explained in 1921, that these boxes, (still used until quite recently) were 'designed to comfortably take a piece of folded cartridge paper of the standard size used for herbarium specimens' and fitted with a 'tin label holder', an idea he 'got … many years ago from a church pew'.[3]

During his decade in 'the tin shed', Maiden welcomed three assistants who became distinguished in their own right—Henry George Smith (1852–1924), Richard Thomas Baker (1854–1941) and Walter Wilson Froggatt (1858–1937)—and established enduring friendships with two other competent botanists, Walter Scott Campbell (1844–1935) who became Chief Inspector of Agriculture, and Henry Deane (1847–1924) the celebrated railway engineer. In 1889, the museum passed from control of the Australian Museum's Committee to the Technical Education Branch of the Department of Public Instruction, and Maiden then became closely associated with the Branch's first Superintendent, Frederick Bridges (1840–1904). The Curator's labours and entreaties were finally recognized in August 1893, when, despite the economic depression, a handsome new Technological Museum was opened in Harris Street, Ultimo. Bridges became Chief Inspector in May 1894, and Maiden served as Superintendent until appointed Director of the Botanic Gardens just two years later.

On a salary of £515, with a residence valued at £105 year, the former Superintendent of Technical Education assumed his new office on 1 June 1896 as yet another great drought was developing. Possessing a width of training, interest and vision not shared by any of his predecessors, he had the additional advantage of nearly fifteen years' experience of the colonial public service, which had sharpened his wits without reducing his energy or enthusiasm.

The new Director clearly supported Lord Acton's contention that there is nothing more important to the man of science than a knowledge of its history. Richard Nichol, employed since 1888 as 'label writer', produced a bundle of old papers which Moore had instructed him to destroy during one of probably many regrettable clean-up operations. Maiden asked Nichol to preserve the papers and promptly had him promoted to 'Label Writer & Museum Attendant'. Nichol later became secretary, and the papers he saved, including letters from Sir Joseph Banks, Baron von Mueller and Sir Henry Parkes, and records of the Botanic Gardens Committee of the 1830s and 1840s, ultimately passed to the custody of the Mitchell Library and Archives Office, where they remain to delight botanical historians.[4]

Maiden's long and remarkably productive term of twenty-eight years as Director of the Gardens may be considered in three phases—the period before Federation, 1896–1900; the pre-war period, 1901–13, and the final decade, 1914–24. His impact upon the Gardens was immediate and dramatic. Having reviewed all the previous Annual Reports he could find, Maiden noted the many years for which no account was given, and in March 1898 presented the first Report on the institution for twenty years, with the declared intention of reporting annually. He also wished Mr Moore 'a long period of leisure after his exceptional services to the Colony'. Annual Reports assumed a new form. They were long, detailed and meticulously prepared, with every facet of responsibility identified and accounted for: correspondence received and answered, costs of improvements, additions to the library, publications produced, seeds and plants received and despatched, herbarium specimens collected and accessioned, museum specimens added, gifts received, amenities provided, amenities required, staff movements and the prevailing condition of the Botanic Gardens, Palace Gardens, Domains, Centennial Park, Campbelltown Nursery and other reserves. There was no question of Mr Maiden's Reports being returned to him for resubmission.

On taking office, Maiden promptly overhauled the medicinal plants garden, and developed the systematic gardens or 'Arrangement Grounds', 'the most directly educational ... of any portion of the Gardens'. The plots of dicotyledonous plants were largely remodelled and the monocotyledons were augmented by 'a fine collection of Australian grasses'. Arrangements were confirmed with Hugh Dixson for the transfer of his collection of Australasian orchids to specially provided accommodation, and attention paid to overcrowded and decrepit plant houses. The ponds and the creek which fed them were cleared of an enormous quantity of silt which was used for top-dressing, and silt-catchment pits were dug to control drainage; 'six iron street orderly bins' of Council pattern, painted green, were procured in the hope that they would 'be increasingly used as visitors learn why they have been erected'; old water-pipes were renewed; cess pits were condemned and sewerage was begun although this required blasting through sandstone to 30 feet and more; old or decayed trees considered unsightly were removed, and a spraying machine was purchased to deal with fungous infections and (on W.W. Froggatt's advice) with insect pests as well. Superintendent George Harwood and his twenty-eight gardeners and the few other staff quickly appreciated that a new Director had arrived.

In the Domain, comprising some 130 acres, some of the 4 miles of roadways and $7\frac{1}{2}$ miles of footpaths received due attention, a wattle plantation was established near Governor Bourke's statue, and an acre of low-lying land opposite Woolloomooloo Bay was filled to make a practice field 'for our young cricketers' for it had to be 'borne in mind that the Outer Domain is the natural play-ground of children who live in Woolloomooloo, many of whom are of poor parents ...'. Maiden rejoiced that Centennial Park was in the charge of William Forsyth and hoped that native vegetation could be preserved there, close to the city where it was still being exterminated.

Maiden declared that men should be required 'to wear suitable bathing costume' for their natatorial activities in the Domain baths, and found that men's latrines, which had 'only recently' (1895) been provided in the Domain were 'on a quiet Sunday' late in 1897, meeting the needs of 1540 men according to a count by the overseer. 'It would prevent much suffering and inconvenience' declared the Director, if latrines were also installed near Mrs Macquarie's Chair, 'and surely, in these days of granting justice to the other sex, it is only necessary to point out that latrines for women and children ... are an imperative necessity'.[5] After all, on a fine Sunday afternoon, 'some 15,000 people may be seen strolling over the grass', reclining under the trees, or listening to diverse discourses from the orators'.

It was also regrettable that there was only one drinking fountain in the whole of the Domain. Another was sorely needed near Mrs Macquarie's Chair, and further, there were in 1897 only thirty-nine gas lamps in the Domain, when the whole area could be more effectively lit by electricity. The Campbelltown Nursery, under John McEwen, despatched nearly 80000 plants during 1897, and was 'a credit to the subdepartment', but clearly it required as an adjunct an arboretum of timber trees for the dual purpose of education and supply. Little wonder the new Director confessed at the end of his fourth year, to being 'still, like Oliver Twist, asking for more'. Unlike his Dickensian model, he was both persuasive and successful, declaring in 1900 that 'Parliament is always generous in its appropriations to the Gardens ...'.

In his first Report, Maiden noted that 'next to the Garden and the other outside establishments ... the care of the Herbarium has been my greatest solicitude'. This concern was shared by Ernst Betche, who was 'practically the keeper of the herbarium', and the Director ensured that parliament also became concerned to the extent of voting funds for a new herbarium building. During his first eighteen months, Maiden, with the support of Ernst Betche, the shy, German-born collector whom he had promoted to Botanical Assistant; William Forsyth, the 'sound botanist' in charge of Centennial Park; and Julius Henry Camfield, overseer of the Palace Gardens, worked feverishly to stock the depleted herbarium he inherited. He lamented that

Opposite: Plan of 1897 which accompanied Maiden's Guide *of 1903. It shows the clear divisions of Upper, Middle and Lower Garden.*

PLAN OF
THE BOTANICAL GARDENS

SCALE OF FEET

Compiled, Drawn and Printed at the Department of Lands, Sydney N.S.W. 18ᵗʰ Janᵞ 1897

Explanatory Notes

○	Represents a tree larger or smaller according to size or importance of tree.	C	Propagating and plant houses.
	Represents Shrubbery	D	Orchid and show houses.
	do special collections of rare plants.	E	Bush house.
	do arrangement grounds for natural orders.	F	Frame and Nursery grounds.
	do artificial rockeries.	G	Carpenter's shop and stable yard.
	do Bamboo clumps.	H	Superintendent's residence.
	do Flower Beds renewed annually.	I	Levy memorial fountain.
	do do with tree in centre.	K	Marble drinking Fountain.
	do Statue or Vase.	L	Refreshment and Ladies' rooms.
	do Drinking Fountain.	M	Monument to Allan Cunningham.
	do Stone or Cement Basins or Fountains.	N	Group of Statuary.
A	Director's Residence and Grounds.	O	Lodges and Residences.
B	Offices, Museum etc.	P	Public Latrines.

For explanation of small Block Letters (A to G) see text.

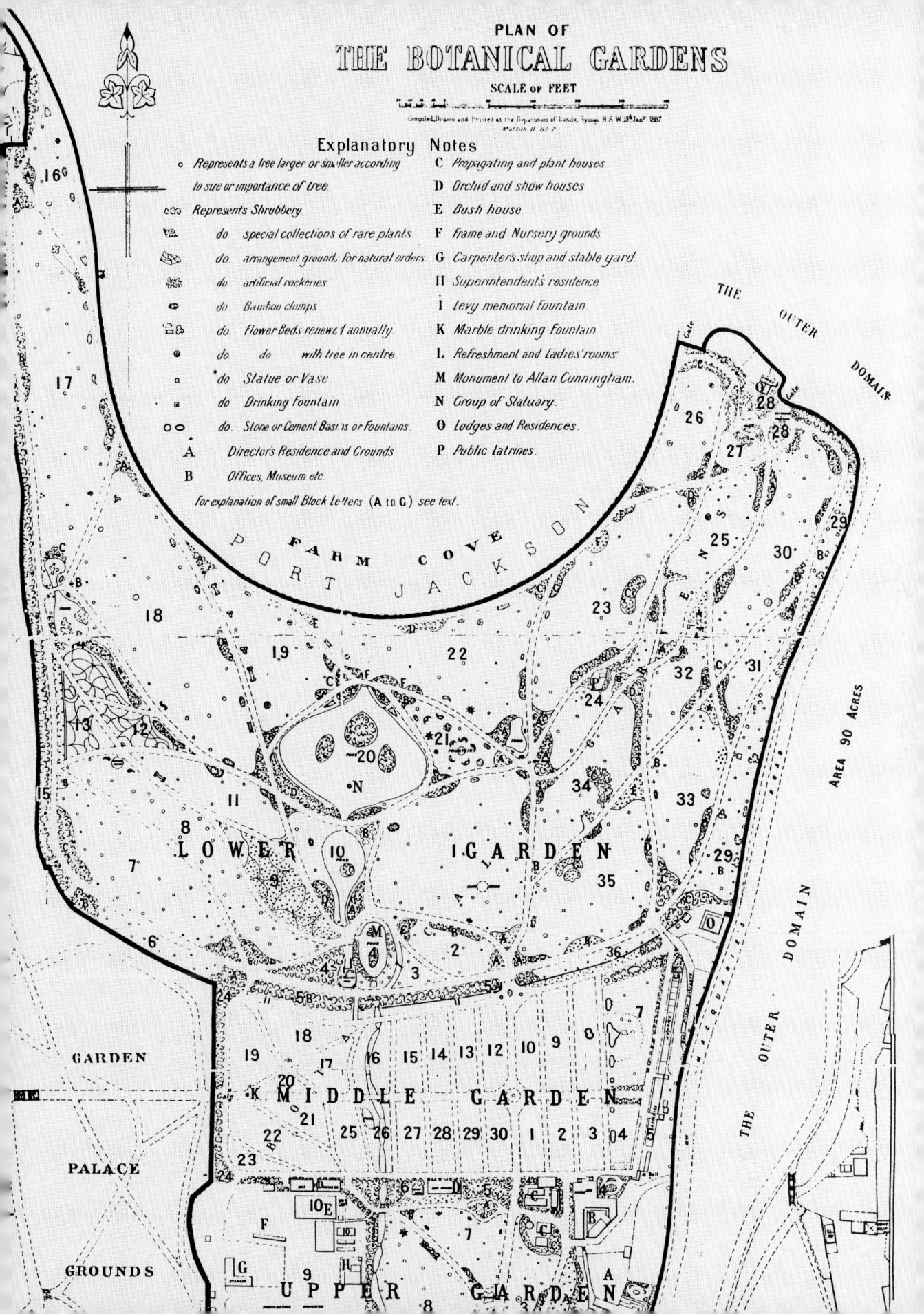

PORT JACKSON FARM COVE

THE OUTER DOMAIN

AREA 90 ACRES

LOWER GARDEN

MIDDLE GARDEN

GARDEN

PALACE

GROUNDS

UPPER GARDEN

THE OUTER DOMAIN

after this period 'of incessant labour, barely half the natural orders are properly arranged'. Maiden himself visited Bourke, Wagga and Cooma before the end of 1896, and during 1897, went to Victoria and South Australia, and to various areas of NSW. He was especially interested in the classic collecting grounds of the early field workers, travelling from Port Macquarie to Walcha in December 1897 through country explored by John Oxley, Charles Fraser and Allan Cunningham; in 1898 to Mount Kosciusko, Jenolan Caves and Lord Howe Island, to Mount Tomah where George Caley, Allan Cunningham, William Woolls and Charles Moore had botanized so successfully, and to Illawarra and the South Coast over some of the collecting grounds of Robert Brown, George Caley, Allan Cunningham and Sir William Macarthur, thereby enriching 'the Herbarium with several thousands of specimens'.

Meanwhile, work continued in the Botanical Museum as 'an essential adjunct to the Herbarium', and the Director began to prepare a detailed but 'inexpensive guide to the Gardens'. He was delighted that the special collections were being frequented by an increasing number of students and hoped to make the Gardens generally more informative for the casual visitor.

Maiden was justifiably proud of the 'commodious building of handsome design' planned (with his constant advice) by the Government Architect to house the herbarium, museum, library, and administrative offices. Built by James M. Pringle of Waverley for £4500, the building was substantially completed by December 1899, after the Director had enjoyed his new office for a month. 'Scientific botany now has its headquarters in New South Wales,' he declared. 'A botanic garden cannot properly perform its functions without the support of a rich herbarium ... a garden of dried plants.' He then promptly asked for 'a picturesque building to serve as a joint laboratory' for various fields of botanical research.

With the able and oft-acknowledged help of Betche and Camfield, Maiden worked principally on the 15 000 and more specimens of flowering plants and ferns. He delegated charge of the mosses to William Forsyth who within two years had 'about 1,200 species'; the fungi to Alexander Grant, the modest, retiring Scottish propagator who had been at the Gardens since 1882, and who in 1899 added 'about 120 glass jars of fungi preserved in the wet way'; and the algae to the beloved science master of Sydney Grammar, Arthur Henry Shakespeare Lucas (1853–1936) who soon had about 750 species in addition to some seaweed specimens collected by the celebrated Professor William Henry Harvey (1811–66). The lichen collection, largely comprising over 20 000 specimens purchased by the government from the Reverend Francis R.M. Wilson (1832–1903), Presbyterian minister of Melbourne, remained without a custodian until after August 1901 when Edwin Cheel (1872–1951) a gardener at Centennial Park since December 1897, transferred to the Gardens. With the appointment in June 1900 of Miss Sarah Hynes as Second Botanical Assistant, to work chiefly on exotic species, Maiden must have felt that he had the staff to develop the new herbarium.

Many 'outside' matters demanded attention. After his time in 'the tin shed', Maiden had an aversion to corrugated iron, and he took pleasure in removing the old and unsightly fence of this material between the Palace Garden and Botanic Gardens proper. He demolished other such fences whenever possible and condemned 'the erection of a hideous fence some 18 feet high' by one Domain baths' lessee who had thereby merely replaced one scandal by another. The floating baths structure in Farm Cove, was judged 'especially hideous'. In 1898 the 'forty-seven feeble gas lamps scattered over a mere 50 acres of the Domain', still deprived citizens of the use of the reserve for evening promenades, and the ladies still waited for facilities affording that relief granted to gentlemen. Nor was the demand likely to decrease, for on 11 December 1898, between 2 and 5 p.m., 11 140 pedestrians, 196 vehicles, 56 bicycles and 7 equestrians were counted as they passed through the St Mary's and Macquarie Street gates to the Domain. The new Director was no dreamy scientist but a practical man who habitually posed awkward questions for those entrusted with public funds ostensibly collected to provide for public welfare and the colony's advancement.

Not only reports and submissions, requests and requisitions poured from Maiden's pen. A constant stream of pamphlets, papers and articles deluged the Government Printer and refreshed the editors of scientific journals and the public press. The fact that his institution had no '*Bulletin* of its own' was cause for only brief regret, for, said Maiden, 'I can address the public on botanical and economic matters in the columns of the *Agricultural Gazette*'—and 'address the public' he did. He also gave public lectures especially to YMCA and church groups, often illustrated by lantern slides prepared from photographs. As Maiden later indicated, the slides had clear advantages:

In the early days of my directorship I used to give public discourses in the Gardens in front of the growing plants. These were so largely attended that the crowds could not help trampling the plants and verges in the vicinity of my stand, and so they had to be discontinued simply because they were successful.[6]

From the outset, an enormous correspondence was maintained with institutions and individuals, the library was greatly enriched, and the plant exchange programme was vigorously developed. The first *List of Seeds available for Exchange* (chiefly NSW and other Australian species) issued in 1897 aroused an overwhelming response. Maiden declared the need for a good rock garden, and in 1899 imported 'over a hundred species and varieties of *Opuntia*' in order to settle the nomenclature of those species of prickly pear declared noxious and to investigate the horticultural potential of others. Maiden 'kept this batch of monsters' growing in the Palace Garden 'in spite of daily protests and gibes, until his retirement', and in 1922 maintained that he had 'been interested in the utilisation of the Prickly Pear rather than the destruction of it'.[7]

Queen Victoria's Diamond Jubilee was commemorated in the Palace Gardens on 22 June 1897, when the Governor, Viscount Hampden, unveiled Signor A. Simonetti's impressive monument to Governor Phillip. Maiden had the water from the fountain diverted to flush Botanic Gardens Creek and replenish the ponds in the Lower Garden. The Jubilee was further marked by thousands of Chinese lanterns and fairy lamps installed with arc lights in the Outer Domain, converting it 'into a veritable fairy scene', while in Centennial Park, military manoeuvres attracted thousands of spectators who 'devastated the vegetation in a manner which, though unavoidable, was distressing to see', for the preservation of natural vegetation was 'a really national work'. Another national work which Maiden strongly advocated from 1897 was a 'Botanical Survey of the Colony', based upon accurate maps, to cover the fields of Pure Botany (including Local Floras), Agriculture, Forestry and Horticulture. In

Maiden's Prickly Pear collection in the Palace Garden.
(Annual Report of the Botanic Gardens, *1909*)

this, Maiden felt that he had 'no more valuable coadjutors than surveyors' such as John Fauna Campbell (1853–1938) and Richard Hind Cambage (1859–1928), both capable botanists.

The Director must have taken comfort in the knowledge that by the end of the century, his beloved Garden, 'Sydney's fairest Paradise', was adequately drained and sewered, the herbarium was being rapidly developed, new show cases were being prepared to display in the museum, not 800 specimens as in Moore's time, but over 7000, and fumigating sheds had been constructed to control plant diseases by hydrocyanic gas. Bands were playing regularly to crowds of one or two thousand, the Woolloomooloo children were enjoying their new playground, and the number of 'domain night-sleepers and habitual loafers' was believed to be diminishing. Furthermore, 'quite a number of gardeners both at the Gardens and Centennial Park' having 'acquired excellent botanical knowledge' seemed likely to heed the Director's advice to join Technical College classes and continue their studies. Regrettably, some handsome trees had been removed reluctantly because 'distant vistas are now desired' at Government House, but there had been pleasing compensation when five acres of the Inner Domain were ceded to the western part of the Gardens, enabling the encirclement of Farm Cove to be virtually complete—and, at last, the Domain was being illuminated between 7 and 10 o'clock nightly, except Sunday, by sixty-eight electric lamps mounted in groups on 60 foot poles.

There was probably some relief in the Principal Secretary's Department when in mid-June 1900, the Director went on leave to attend the International Botanical Congress in Paris, and to inspect botanic gardens, parks and herbaria in the United Kingdom and Continental countries. Earlier in the year he had gone on collecting trips to the Victorian Alps, the Wyong District and the Blue Mountains. Deeply moved by the staff's 'kind words' and presentation of 'a beautifully fitted travelling bag' which clearly indicated 'the very cordial relations that exist between the staff and myself', Maiden left for Europe. George Harwood took over for six months.

Maiden resumed duty in 1901 thankful for his good fortune 'in possessing a loyal, intelligent, and hard-working staff'. Harwood had arranged the transfer of William Faris Blakely (1875–1941), gardener at Jenolan Caves, and in 1901 he was appointed to the Gardens on probation at 7 shillings a day. He quickly proved his worth, and in June 1907 was commended for 'a highly meritorious list of Economic plants growing in the garden, together with suggestions for guides to the garden'. By this time, Maiden's long-promised sixpenny *Guide* which had 'involved an enormous amount of labour on the part of myself and some members of my staff', had been in use for four years. It remains a valuable historical record.

The years 1900–01 brought the Gardens and associated reserves into considerable prominence. On 23 April 1900 '25,000 people flocked to the Outer Domain, to view the departure of the Imperial Bushman troops for ... South Africa'. This caused 'rather heavy wear and tear, but very little wilful damage'. On 15 December Lord Hopetoun, the first Governor-General, landed at Farm Cove, and the opportunity was eagerly taken to tow 'Mr Cavill's floating baths' around to Woolloomooloo Bay, and to replace them with 'a floating landing stage ... decorated in Venetian style'. In the same month, a military camp was established in Centennial Park in association with the Federation festivities, and on his return, Maiden expressed the hope that 'this will not be repeated'.

Processions to mark the inauguration of the Commonwealth in January 1901, and the visit of the Duke and Duchess of Cornwall and York in May, both began in the Outer Domain, and during the year the demand for space on Sunday afternoons increased. Twice an Anglican bishop 'in his canonical robes' addressed 'large and well-behaved audiences', while elsewhere bands played, often within earshot of each other, and zealots of Protestantism, Catholicism, atheism, phrenology, certain political policies, and 'stump orators of various kinds' expounded their views. One George Perry, 'temperance lecturer and total abstinence advocate' had been 'twenty-three years in the Domain', and scarcely a Sunday passed that did not feature speakers and sometimes musicians, from: Church of England (St James's, Sydney), Church of England Mission (St Peter's, Woolloomooloo), Central Methodist Mission (Centenary Hall, York Street), Baptist Mission (Burton Street, Woolloomooloo), Salvation Army (which braves all weathers), Plymouth Brethren, Christian Israelites, and Christadelphians. For variety, there were health lecturers, blackboard artists and 'lightning calculators'. The 'safety valve' tradition has been maintained. While the adult 'orators and listeners ... alike good-humoured' charmed each other, the children tended to indulge their desire to climb 'trees, posts, rails, and fences'. It was therefore decided 'to fix up a juvenile gymnasium ... fitted with swings, trapezes, seesaws, parallel and horizontal bars, and ladders of rope and wood'. This children's paradise on the site of the subsequent chess tables was declared to 'be an innovation for Australia' and it proved both remarkably successful and surprisingly safe.

The 'juvenile gymnasium' in the Domain.
(Annual Report of the Botanic Gardens, *1907*)

Centennial Park was the site of the swearing-in ceremony of Lord Hopetoun as first Governor-General, and it was 'estimated that about a quarter of a million people were present' on 1 January 1901, yet there was less damage to the Park 'than sometimes takes place at the usual annual military review'.

While 'not in favour of maps, diagrams, and profiles being introduced into carpet bedding as a rule', Maiden was happy to acknowledge the efforts in his absence of 'Messrs Allen and Lovegrove, who have charge of the carpet bedding', and who had created a new carpet bed in the Lower Garden, depicting a map of Australia showing the names of the federating colonies. This was 'warmly commended by the public' and the Director conceded that it was 'altogether exceptional', but he showed rather more enthusiasm for the tree plantings performed by the Royal visitors in the area recently acquired from the Inner Domain. Other 'bedding-out arrangements' to mark Federation greeted visitors entering the Palace Gardens by the gates opposite Governor Bourke's statue near Macquarie Street.

Part of the old Museum and Lecture Hall (1878) was incorporated into the National Herbarium, Museum and Library built in 1899–1900 and opened in 1901.
(Royal Botanic Gardens Library)

On 8 March 1901 the Colonial Secretary, John See, and his wife, officially opened the new Botanical Museum and National Herbarium, and it was noted that the latter contained all but 'very few' of the described plant species of the State. To maintain this desirable situation and to develop his notion of a botanical survey, Maiden promoted John Luke Boorman in October from gardener to collector, a position he retained until after the Director retired. Another very significant appointment was that of Miss Margaret Flockton (1861–1953) who commenced duty on 3 June 1901 as 'artist' at 2 shillings an hour, later fixed at 10 shillings a day. She remained at the Gardens until 24 March 1927, leaving as her monumental contribution the painstakingly accurate plates with which the Director was able to adorn his two great works, *A Critical Revision of the Genus Eucalyptus* and *The Forest Flora of New South Wales*. These began to appear in parts, respectively in 1903 and 1904, and continued for twenty years and more. Margaret Flockton also collected specimens, illustrated species of *Opuntia*, designed Maiden's bookplate, and gave tuition to his second daughter,

Mary, who about 1905 also produced illustrations of considerable merit (signed 'M.M.') for her father's publications. 'An artist is an indispensable officer … of a botanic garden', declared the Director, adding with a fine historical flourish that 'Sir Joseph Banks long ago' held the same opinion.

During this second phase, 1901–13, Maiden gave full rein to his predilection for history. Following the enjoyment of showing his children places 'of antiquity and stability' in England, there was the excitement of returning to the historic Commonwealth celebrations and a Royal visit. In London, Maiden called on James Britten (1846–1925) of the Botany Department, British Museum, and 'begged for a few duplicates' from the Australian collections of Banks and Solander. Britten replied: 'I will give you a few with pleasure, when we distribute them.' The delighted Director 'was quite content to wait five years for the fulfilment of the promise', and was even more delighted when the 'few' proved to be 586 species, by far 'the most valuable acquisition' to the Herbarium in 1905.[8] It was a far cry from the time in 1876, when, recorded Maiden, 'Sydney was offered one of the duplicate sets' of specimens collected by Robert Brown, but unaccountably failed to respond.[9]

Immediately after the Federation festivities, work began in earnest on clearing the old Devonshire Street Cemetery to make way for Central Railway Station, and Maiden ensured that the remains of his eminent predecessor, Allan Cunningham, would not be consigned to oblivion. The few bone fragments recovered from the grave 'were placed in a leaden casket, 10in. long, 5in. broad, and 4 in. deep' and on 26 June 1901 reverently deposited in a cavity cut into the memorial obelisk which had been erected in 1844 near the site of the present restaurant. Unfortunately, the graves of other early superintendents, James Anderson and Nasmith Robertson in the same cemetery, did not enjoy similar protection, and were obliterated some seventy years after their transfer to Botany in 1901.

By this time Maiden's research had equipped him to lecture and to write extensively on the history of the Gardens, the careers of his predecessors and the work of botanical pioneers in each State. Readers of the *Agricultural Gazette of NSW* (1902), the *Public Service Journal* (1902–04) and the proceedings of learned societies and conferences, including the Australasian Association for the Advancement of Science (1907, 1909, 1911) were treated to a feast of articles demonstrating the importance of the human side of scientific pursuits. In 1906 Maiden's historical notes on the Gardens, written for the *Sydney Morning Herald*, were republished in the *Kew Bulletin*. The 'large mass of manuscript' still unpublished at his death, appropriately went to his old friends, R.H. Cambage and W.W. Froggatt, for editing and presentation in 1927 and 1931 to the Royal Australian Historical Society of which Maiden had been an early president and vice president. Maiden's knowledge of history had long been widely appreciated when he was consulted on botanico-historical matters by such workers as Frank M. Bladen and Frank Walker, two of the founders of the (Royal) Australian Historical Society. As secretary of the 'Banks Memorial Fund', Maiden in 1909 published *Sir Joseph Banks: the Father of Australia*, to commemorate 'a great and a good man, worthy to be honoured by every Briton whether Australian or not'. Proceeds were intended for 'a statue of Banks in the vestibule of the Mitchell Library.'[10] As Director of the Gardens, he established a portrait gallery of his predecessors and of

View of the Gardens about 1900 from the corner of Bent and Macquarie Streets.
(Mitchell Library, Sydney)

the great botanists of the past, to adorn the walls of the museum and herbarium with appropriate reminders of purposeful endeavour.

The early years of the new century were not only productive from the scientific, historical and literary viewpoints. Gas light remained unchallenged until 20 August 1908, when on the eve of the landing of 3000 men from the visiting American Fleet, forty-nine gas lamps in the southern Domain, costing £4 per lamp per year, were supplemented by thirty-two 'electric flame arc lights' suspended from 30 foot poles in the northern area, costing £23 per lamp per year. In 1910 thirty-nine of the remaining gas lamps were 'superseded by thirty-three Excello arc flame electric lamps of alternating current, fixed upon tall, iron standards' but the main entrances remained gas-illuminated.

As one problem was solved another arose, and the Director noted in 1902, as 'a sign of the times ... the large number of these horseless carriages now to be seen in the Domain'. Although they did not appear to have any undue effect on horse traffic, he considered that 'petrol emits an unpleasant odour, which does not punish the automobilist, but which is nauseous to pedestrians and others in his rear' and the raising of dust was also 'disagreeable to others'. In 1903 the automobiles met a formidable competitor when a 10-ton steam-roller was used for the first time 'to thoroughly consolidate the blue metal' on the Domain roads. A speed limit of 8 miles per hour was imposed, and in October 1908 four motorists were 'fined £3 each and costs for reckless driving in the Domain'; a motor cyclist's similar behaviour cost him £2. The Director probably wished that more visitors would quietly use the 400 seats he had dutifully installed for their comfort.

Clearly the Domain was rapidly changing in many ways, and guided by his historical awareness, Maiden enlisted the aid of Julius Camfield in 1902 to compile a census of 'plants growing without cultivation', both indigenous and exotic, in the Outer Domain, 'for the information of the present generation and of posterity'. This long and carefully composed list stirred the memories of people then living who remembered the Domain as 'a place of wild flowers'.

Even in the comparative tranquillity of the Gardens, 'these noisier times' were acknowledged, and the old bell, dated 1839, rung at intervals during the day, had proved too small. It was despatched to the Campbelltown Nursery, and a larger bell began tolling on 13 October 1902.

The cession of 5 acres from the Inner Domain to the Gardens proved something of a mixed blessing, for between 1900 and 1905, over 200 000 cart-loads of filling from city building sites were used in regrading 10 to 12 acres of the Lower Garden to produce a harmonious and practical alignment with the new area. Worthy projects took their toll, as on 24 April 1906, when opposite the Palace Garden work began 'in connection with the building of the "Mitchell Library", on the site of the wattle-tree plantation' which then boasted '120 fairly well-grown' young wattles representing seventy-eight species. Then there was the telephone, which the Director considered was used 'too freely' by citizens seeking information about plants. He preferred written enquiries to which, after due reflection, he could supply written replies. Thus records could be kept (frequently 5000 to 6000 letters a year were received) and misunderstandings and misinterpretations avoided.

Palm bed and rockery (with an array of labels) in the Lower Garden about the turn of the century.
(Royal Botanic Gardens Library)

The census of Domain plants may well have provided welcome relief from work on the census of the flowering plants and ferns of the whole State, on which Maiden and Betche had then 'for some years been engaged'. The *Census* was not published until 1916, the first since Moore's list of 1884. Apart from the literature, there was abundant material on hand. In 1903 Betche estimated that the herbarium contained 97 per cent of all described species in New South Wales; 96 per cent of Victorian species; 93 per cent of Tasmanian species; 85 per cent of species in 'extra-tropical South Australia'; 62 per cent of Queensland species; 46 per cent of North Australian species, and 36 per cent of West Australian species. There were in addition 30 000 exotic species. Maiden felt that '... all the Australian States, except perhaps Tasmania and Victoria, are botanically so imperfectly explored, that in some cases 10 per cent, and more will be added in the future'. About midnight on 6 January 1904, it seemed for a while that all would be lost, when the offices, herbarium and museum were threatened by a fire which gutted a nearby potting shed and stokehole.

Maiden continued to participate in the botanical exploration he advocated, but on one excursion while collecting orchids for the Gardens and material for the *Forest Flora*, he had an accident. A leap 'across a small ravine at the top of Macquarie Pass' caused a serious hernia, which with the aid of a truss, he endured for a few years before surgery in 1911. Thereafter he became increasingly lame, a condition aggravated by rheumatoid arthritis. Another setback came with the news that his only son, Harrie, a twenty-year old merchant seaman, had been lost on 31 January 1905 when the barque *Eulomene* sank in the North Sea 'and neither he nor Mrs Maiden ever quite recovered from the shock' although work proceeded, apparently unabated.[11]

Charles Moore died three months after Harrie Maiden, and the Director reviewed his predecessor's long career, noting also that the Gardens now contained 'more native plants than at any previous period' and that people were showing 'a steady desire to know more of their own beautiful and interesting flora'. In the same year, Edwin Cheel went at his own expense to England where he learned more about lichens. After Alexander Grant's death on Christmas Day 1906, he was given the additional charge of the fungi. Other changes in the work of the herbarium shortly occurred when George Israel Playfair (c. 1870–1922) assumed responsibility for the fresh-water algae, and first A.E. Goddard, then the Reverend William Walter Watts (1856–1902) for the mosses and liverworts.

In 1907, following his historical interest, Maiden listed the 'memorial trees' planted in the Gardens since 1868. He noted with relief that the Mitchell Library building had been largely constructed, and with chagrin that the Agricultural Hall of 1879, 'the tin shed', still stood as 'a great eyesore in the Domain'. Another historic event was the transfer of the Gardens from the jurisdiction of the Chief (formerly Colonial) Secretary's Department (to which it had been attached since 1880) to that of the Department of Agriculture as from January 1908, an association destined to last over seventy years. Other points of interest and concern were the capture of large native cats (*Dasyurus*) in the Botanic Gardens, the sighting of rabbits in the Palace Gardens, the constant need to impound cattle and horses in Centennial Park (where flower thieves with a preference for stocks and chrysanthemums habitually eluded apprehension) and the tendency for large fig trees to become over-possessive, while 'the old primaeval Eucalypts' continued to die.

The 'Arrangement Ground' in the Lower Garden where plants were grouped according to their affinities.
(Annual Report of the Botanic Gardens, *1910*)

In 1910, when the Royal visitors of 1901 became King George V and Queen Mary, Maiden recorded a bountiful year for 'gardening operations', and from 1 July the garden labourers were motivated by an increase of sixpence in their minimum daily rate, now 7/6. Wattle Day, which Maiden fervently promoted, was duly celebrated on 1 September (later changed to 1 August) just a fortnight before the unexpected death of William Forsyth, BA, the quiet, studious overseer of Centennial Park since 1891, and curator of mosses and liverworts until 1903.

The most harrowing event of the year occurred in July, when Maiden 'brought Miss Sarah Hynes, BA ... before the Public Service Board on certain serious charges' involving disposal of 'a basketful' of original labels from Henry Deane's eucalyptus specimens, then being held on trust in the herbarium. Miss Hynes countercharged that Maiden had destroyed the labels. The mortified Director assured Deane, 'I have not destroyed one of your labels in my life & no member of my staff dare destroy an original label'. He asked Deane to come '& examine *every* specimen'.[12] The luckless Miss Hynes was found guilty of insubordination and disobedience, fined, and transferred to the Department of Public Instruction to teach botany.

Among happier events of 1910 was the construction near the aviary of an 'Insectarium' where the Government Entomologist, W.W. Froggatt, could study 'the life histories of plant pests'. The aviary itself continued to provide delights for visitors and dilemmas for staff during Maiden's time. One of his first acts was to establish a 'Register of Birds' in a large volume used until October 1940. As for seeds and plants, the purchases, presentations and exchanges of birds were recorded. The range of birds, indigenous and exotic, was quite comprehensive—blackbirds, canaries, cranes, cockatoos, ducks, emus, eagles, finches, galahs, hawks, hornbills, jackdaws, kiwis, kookaburras, lyre-birds, magpies, ostriches, pigeons, penguins, pheasants, swans, turkeys, and so on.

Additions to this ornithological company included a turtle, tortoise, 'opossum', 'porcupine' and a hive of Italian bees. Perhaps the tortoise was that large quiet denizen of undisturbed places which long remained to remind observant visitors that there had once been a zoo in the Gardens. She died, apparently of old age, about 1967 when the then Director, Knowles Mair, conveyed the revered reptile to the Australian Museum to be accorded the services of a taxidermist. Fates of many birds before the removal of the aviary about 1940 were sometimes rather more violent. Some simply escaped, were let out or were stolen, but others were believed to have died from extreme heat, to have drowned in water troughs, or to have been killed by worms, ants, rats and cats. One black swan was 'killed by geese', another 'by a dog', and three others were 'hit on head with blunt instrument'. Some, like a golden pheasant cock, died fighting, while others fell victims of a hawk which went berserk. An emu apparently 'killed itself', a goose died after a inquisitive or vicious visitor poked a stick into its eye, and two birds died of 'natural causes', presumably fright, when 'disturbed at night by rat catchers'. Having avoided such disasters, and the attentions of the native cat found nearby, some birds fared long, if not happily, in captivity. In 1911 a brolga or native companion died after twenty-five years in the aviary; and in 1912 a cockatoo from Northern Australia died after living 'in the Aviary nearly 50 years'. In May 1915 a kookaburra was presented to Colonel Thomas Fiaschi 'as a mascot' for the 3rd General Hospital, AIF.[13] Further zoological activities were undertaken in 1913 and 1914 when specialists including William Joseph Rainbow, Anthony Musgrave, Emil Zeck and Alfred J. North compiled checklists of the Gardens' fauna, including insects and arachnids, molluscs and batrachians, fish and reptiles, birds and mammals.

During the years immediately before the Great War, there were significant staff changes. In January 1911 Arthur Andrew Hamilton was promoted from Gardener, Centennial Park to Botanical Assistant, Botanic Gardens; additional links with the Department of Agriculture were established through the transfer to the Gardens first, in May 1911, of Ewen Mackinnon, who doubtless to Maiden's delight, required improvised 'Laboratory accommodation', and second, in June 1913, of Ernest Breakwell (b. 1884) as agrostologist. Regrettably there were also some losses. George Israel Playfair, 'almost a solitary worker' in NSW on freshwater algae left Sydney in August 1912; on 30 January 1913, John McEwen died, having been superintendent of the Campbelltown Nursery, 1884–1912; on 28 June 1913, Ernst Betche died, 'a fine character, conscientious, hard-working, unselfish, just', and was succeeded by Edwin Cheel as Senior Botanical Assistant; in the following month, James Jones retired, having served successively at Wentworth Park, Centennial Park and the Domains since 1884. Some of his improvised survey instruments were presented to the

The north-east corner of the old Herbarium about 1910. This room is now part of the Anderson Building and houses the secretarial staff.
(Royal Botanic Gardens Library)

museum. Jones's foreman, John Washington (b. 1840) retired at about the same time after thirty-two years' service. The practical Director greatly admired Washington for

he was tactful with his men, and not only told them what to do, but showed them how to do it. He was a most industrious man, and a perfect artist with the shovel. I shall always remember his skill in putting a crown on a piece of roadway, or in distributing screenings. It was a pleasure to see his good work.

Literally on the brighter side, the whole of the Domain was lit by sixty-five electric lamps in 1911, and the outdoor staff now worked 7:30 a.m. to 5 p.m., Monday to Friday and 7:30 to 11:45 a.m. Saturday, being no longer required to start work at 6 a.m. during the warmer half of the year and at 7 a.m. during the other half. Within two years, the minimum wage for general labourers was 9 shillings as day, for gardeners, 10 shillings a day, and for propagators, 11 shillings a day.

Men and horses pose for the camera during landscaping work in 1912.
(Royal Botanic Gardens Library)

In presenting his report for 1912, Maiden gave notice that 13 June 1916 would mark the centenary of the Botanic Gardens, and in providing evidence for this, suggested that it be celebrated 'in a suitable manner'. On the more immediate and practical side, 'the idea ... occurred' to the Director, '... of converting some of our old-fashioned fountains into cupless ones A local plumber has been employed ... to convert the central freestone, bibcock, three-tap fountain into a sanitary, bubbling, crystal stream three-jet fount—the first in Sydney ...' Thus chains and unhygienic cups began to disappear from the Gardens and Domains and the 'New "Bubble" Fountains' became standard installations.

A disaster loomed in March 1912 when T.E. Keele, one of the Commissioners of the Harbour Trust presented the government with a scheme aimed at reducing shipping congestion by diverting mail steamers from Circular Quay to Farm Cove. As a

Herald reporter put it, 'In other words, the Botanic Gardens of Sydney are to be wiped out, and the Domain is to be converted into a centre of commercial activity!' The normally reticent George Harwood was incensed: 'I call it a wildcat scheme, and I don't mind if you print that' he declared. Warming to the subject, he further declared that 'it would be wicked, it would be a sacrilege, it would be an outrage and a crime to take the beautiful gardens and make another Darling Harbour of them!... There will be a regular uprising. Now, you mark my words!'[14] The *Herald* gently noted, 'the Government has judiciously suppressed the scheme ...', and the hearts of the Director and his staff beat normally once more, although there was some concern about the upheaval caused by 'the widening of Macquarie-street to 80 feet'.

One immediate effect of the Great War, 1914–18, was the reduction of staff through enlistments. Four gardeners, a carter, a clerk and five 'scientific cadets' enlisted in 1915, as did members of the Domains and Centennial Park staffs. On 3 February 1917 the Director unveiled a cedar Roll of Honour, made by the Gardens carpenter, decorated by Miss Flockton, and bearing nineteen names. Enlistments continued during 1917 and 1918. Two gardeners, William Ryan, who had lost both feet, and Charles Clark, who had been wounded, both returned in 1918 and heroically resumed duty. Functions to mark departures and returns were organized by the Gardens Patriotic Committee, which also maintained correspondence with those at the war.

During 1914 Maiden continued his assault on corrugated iron structures, and rejoiced in the disappearance, at last, of the tin shed where he began work in 1882. The 'ground set free by the demolition of this building' was promptly planted with eucalypts. George Harwood, the shy, competent Gardens Superintendent, went on retirement leave on 1 August 1914, and was succeeded by Edward Naunton Ward (b. 1871). Harwood had 'been more concerned in development of the Botanic Gardens than any other man since its foundation, with the single exception of the late Mr Charles Moore ...'. A native of Taunton in Somerset, Harwood had joined the staff on 1 September 1873 to work in the hot houses and subsequently 'in every branch of the establishment'. Maiden confessed: 'I often shuddered lest he should be taken away from me in the early years of my Directorship', and regretted that, as in the case of Charles Moore, he had failed to persuade Harwood to record his reminiscences before he died on 18 January 1915.

Amid rising prices, there was even more need for economy during the war, and there were other problems, for example, the movement late in 1915 by a group of worthy and influential ladies to establish 'a Tea Garden for the use of soldiers'. With some political support, the ladies sought three-quarters of an acre, perhaps behind the Mint and Registrar-General's Office or in the cricketers' area, to be enclosed with 'fencing sufficiently high to prevent men jumping over'. Within this sanctuary a 'Tea Gardens and Cinematograph Centre' would be established. Maiden agreed that the soldiers should have the opportunity to drink tea rather than more deleterious beverages, but suggested that Hyde Park was as 'near the ideal as possible', unless the ladies should wisely choose to 'hire a picture palace and a tea room in the city' for their laudable purpose.

Part of the Gardens Nursery in 1914, showing the bush house, propagating houses and frames.
(Annual Report of the Botanic Gardens, *1914*)

The soldiers' amenity was eventually established near the St Mary's Road entrance to the Domain and early in June 1916 was opened by the Chief Justice, Sir William Cullen. The ladies wanted potted palms for the opening, a key to the Domain gates, a connection to the gas supply …, then the secretary with whom Maiden had had strong differences suffered a breakdown, and other committee members resigned. However, the watchful William Grant reported on 5 July 1916 'that a large, flaring red calico sign … a hideous feature in the landscape' had been raised by the triumphant ladies who thus declared the existence of the 'Anzac Buffet' and infringed 'regulation No.2 of the Gov. Domains'. Ordered to strike their colours, the ladies reluctantly did so, and thus ended the Director's private war to preserve the Domain from such a worthy invasion. He proudly reported that during 1918 'about two hundred' soldiers had visited the Buffet daily, and that it had served as a rendezvous for 24,000 returning soldiers and 'fully one hundred thousand relatives'. On 25 March 1920, Maiden, understanding that the amenity would close in April, wrote to the ladies congratulating them on their 'successful and self-sacrificing labours' and he was courteously thanked for his co-operation.

The 'Anzac Bed' in the Lower Garden. An example of carpet-bedding in 1916.
(Royal Botanic Gardens Library)

Meanwhile, the Woolloomooloo Branch of the Sydney Day Nursery Association applied to use the Buffet for temporary accommodation. Although holidaying at Blackheath, Maiden wrote to George Valder, Under Secretary for Agriculture, expressing the hope that this was a mere rumour. 'I reply with Voltaire that there is nothing so permanent as that which is provisional' declared the Director, adding, 'Surely, my views should be called for before a decision is given'.

A decision was given, despite Maiden's protests that once again he was being kept in the dark about ultimate intentions. The Association took possession of the Buffet in May 1920, and on 18 August, J.J.G. McGirr, then designated Minister for Public Health and Motherhood, officiated at the opening. In September 1921, when the Day Nursery's term had almost ended, the premises were sought by the National Association for the Prevention and Cure of Consumption 'as an Anti-Tuberculosis Dispensary'. Maiden considered the proposal 'a monstrous one'. He believed that 'it would soon get abroad ... that Tuberculosis has become such a scourge in New South Wales, that the Government has had to enclose a portion of Sydney's most important park, close to the legislative buildings and public offices' in order to cope with the problem. On the other hand, 'the claim of the Outer Domain to have the land restored to it (taken for temporary war purposes) is unquestionable'.[15]

The Director won this time. The nursery vacated the building on 7 October 1921 and Maiden and Grant had it down within four days, declaring that the timber would replenish stocks and that the area would be restored 'at the earliest possible moment'.

Within two years 'a convenience for women' enclosed by an iron railing fence and a hedge 'to ensure privacy' occupied the site.

There were other changes and upheavals during and immediately after the war, although work in the herbarium proceeded quietly and effectively under the direction of Edwin Cheel and his assistants, A.A. Hamilton and W.F. Blakely. Hamilton reported on such new fields as teratology (the scientific study of biological variants and monstrosities), ecology and leaf variation, while Blakely became more absorbed in the study of eucalypts; Thomas Whitelegge (1850–1927) assumed charge of the ferns, mosses and liverworts in 1917; Ernest Breakwell continued to work diligently on the seed testing of fodder plants and on other pastoral and agricultural projects, and A.J. North added two birds to the fauna checklist. Work proceeded on the widening of Macquarie Street by 20 feet in 1915 'at the expense of Government House Grounds'; rifle clubs were permitted to use part of the Inner Domain for drill 'in defence of the Empire', and the Australian Museum generously transferred 'the whole of its botanical library' to the delight of the Director and of the 'shorthand-writer and Librarian', Miss A.M. Jenner, 'a model librarian', whose death on 23 February 1918 was widely regretted.

In May 1916, there was the heaviest invasion of flying foxes since 1900, followed on 13 June by a lighter and more welcome invasion of 'a large gathering' of citizens, who despite the soaking wet lawns and dull, threatening sky assembled 'between the Conservatorium of Music and Government House' to observe the centenary of the Gardens. The chairman of the function, William C. Grahame, Minister for Agriculture, affirmed the belief that the Gardens 'were ... the best situated and the most beautiful in the Southern Hemisphere'; the Director gave a brief historical address, pointing out that the Gardens belonged not only to Sydney but also to the nation, and that in another century the descendants of those present 'would see more beautiful Gardens, just as they would see a cleaner and more beautiful city'. The Premier, William A. Holman, considered that the Gardens 'had a treble claim upon the admiration of the people of the State—their artistic beauty, the fact that they were the starting point of botanic investigation as applied to agriculture and horticulture, and that they constituted an extraordinary historical link with the past'. The Governor, Sir Gerald Strickland took the opportunity to refer to the King's recent bestowal upon the Director of the Imperial Service Order.

Fifteen trees of geographically appropriate species were then planted by 'gentlemen representing various dominions and nations in the historical ground' near a new rose garden. The plan, and some of the trees, still exist. Finally, 'the visitors proceeded to a site close to the aviary and national herbarium, where the foundation-stone of the proposed museum of botany and agriculture was laid' by the Minister, and the Director was presented with his portrait painted by Norman Carter.[16] For Maiden, this was doubtless the most gratifying day of the war years.

There were other matters of note. In the Domain, the old carbon arc lights were converted to 'Watt Incandescent Arc' lamps, and somewhat unexpectedly, the Campbelltown Nursery had two record years in 1915 and 1916, despatching over 115000 plants annually to schools, churches, cemeteries, parks, reserves, railway stations, hospitals, military camps, other nurseries, the Federal capital, and of course, to the Botanic Gardens. Then on 26 November 1916, that 'very soul of loyalty', Julius

Henry Camfield died, just five months after Maiden had paid a public tribute to him at the centenary function. A native of Islington, London, Camfield arrived in Sydney in January 1882 and Moore appointed him overseer of the Palace Gardens. Like several of his colleagues, he was 'very shy and reserved', and happiest with his family, plants and books. A sound botanist and 'singularly well-informed', he retired in October 1916 as overseer of the Inner Domain. Maiden's keen sense of loss was shared by Camfield's family, including the granddaughter who long remembered his gentle instruction not to touch the rolls of newspapers hanging by day on the iron picket fence of his Gardens residence. These were the 'blankets' of 'the poor men who slept in the Domain at night'.[17]

In May 1917, the Botanic Gardens Department Progressive Society was formed 'to encourage and assist in the study of all branches of Horticulture' and, of thirty who joined, about twenty attended the monthly meetings. Maiden strongly supported any movement likely to result in a more enlightened staff. Today he would be an advocate of 'job satisfaction' and 'in-service training'.

The aftermath of the war assumed many forms. There were the sad offerings of seeds collected in the deserts of the Middle East, on the battlefields of the Somme and from plants blooming on the graves of fallen comrades. Maiden noted that although such donations 'promise no scientific value, they are being dealt with in a kindly way'. In 1920 and 1921, more seed from France and Belgium was received, some with 'the deepest gratitude and loving sympathy of the school children of Villers Bretonneux'. There was an enthusiastic demand for this seed, chiefly of 'the popular old world cornfield poppy ... the well-known "Shirley Poppy".' By 1923, requests were being received from councils for trees to serve as soldiers' memorials.

By February 1919, Maiden, like other Sydney residents, was 'taking the [pneumonic] influenza very seriously ... and ... travelling as little as possible',[18] but on Saturday, 19 July, a marine display was held in Farm Cove as part of the Peace Celebrations, and

people flocked to the Gardens in numbers never before witnessed by the oldest hands. After the procession, it was difficult to find lawn space to spread a small table-cloth, such is the popularity of the Garden on public holidays. It would be difficult to describe the state of the lawns, paths, and borders at daylight on the Sunday morning of 20th July.

But, 'the occasion was a huge and happy picnic' and the damage to plants was not really serious, although '15,000 turfs were purchased to repair the worst of the damage'. Similarly, the Domain had become 'one huge picnic ground'. It was a relief to celebrate 'the ending of one of the most unhappy periods in the world's history'.

In 1915 provision had been made for a 'motor mower house', an amenity which marked a welcome change from the days of swinging scythes and clattering horse-drawn mowers. Four years later Maiden reported that *Paspalum dilatatum* (ironically, introduced by his respected friend Baron von Mueller in 1891 as a fodder plant) had become 'our worst lawn pest'. He also noted that stable manure was becoming difficult to obtain with 'the ever-increasing use of the motor-car and motor traction'.

During 1920 there was another flying fox invasion which was brought to the attention of the Sydney Gun Club; Major Anthony Hordern of Darling Point presented a fine collection of orchids; the Reverend W.W. Watts died in Melbourne and Maiden

Margaret L. Flockton who worked as botanical artist at the Gardens, 1910–27.
Her work graced many of Maiden's publications.
(Royal Botanic Gardens Library)

managed to purchase his collection 'at a very cheap rate' after 'battling with the Department'.[19] The spades used by the Duke and Duchess of York for tree planting in 1901, were restored to service for the Prince of Wales during his visit in June. At the beginning of 1921 Robert Henry Anderson, a young graduate in Agricultural Science, commenced duty as Botanical Assistant on six months' probation at £294 year and was soon at work in the museum, the herbarium and in the field.

By early 1922 Maiden was in correspondence with G.P. Darnell-Smith about eucalypts. Darnell-Smith, a biologist with the Department of Agriculture, was especially interested in galls on eucalypts, and he confessed that in Maiden's *Critical Revision* 'the portions dealing with ecology and physiology are perhaps more interesting to me because I know more about them than the purely systematic work'. Maiden himself was disappointed that the *Critical Revision* was being printed 'slowly because of the financial depression'.

In March, 'a request for flowers for decorative purposes' at a hospital ball found the Director at his responsive best: 'This is not a flower farm, and flowers are not grown for the purpose of cutting ... they are not grown in masses like tradespeople grow them ...'.[20] Furthermore an undesirable precedent would be set, the staff would be discouraged and the Department would fall 'into bad repute with the floral trade ...'. By contrast, there was the news that Major Hordern planned to donate the remainder of his orchids, 113 plants, chiefly *Cattleyas*. In May 1922 the Royal Society of St George sought action to preserve the lettering on Mrs Macquarie's Chair. Maiden took steps to prevent erosion but declined to tamper with the original lettering which was simply lined with black paint under William Grant's supervision.

In his twenty-eighth and last Annual Report of 1923, Maiden condemned the false economy which obliged his herbarium workers to use newspaper instead of 'the herbarium paper hitherto used'. Though reduced by wartime restrictions, Maiden's Reports had been models of informative composition. The first Report issued after his departure was a fifteen-line paragraph of staccato sentences in the midst of the thirty-four-page Annual Report of the Department of Agriculture. One of them read: 'As previously stated, Mr J.H. Maiden, I.S.O., F.R.S., F.L.S., ceased active duty as Director on 25th April and went on extended leave prior to retirement'.

Happily, the introduction to the Report was more generous with words and praise, and tribute was paid to Maiden's 'services ... of inestimable value', to his 'botanical works' which would 'stand as an everlasting monument to his name', and to the fact that his 'heart and soul were wrapt in his work'. Meanwhile, the Director left the Gardens residence where he and his family had lived so long and happily, and moved to 'Levenshulme', in Turramurra Avenue, Turramurra.

On Thursday, 2 October 1924, Maiden, who now walked 'with great difficulty and much pain',[21] returned to the Gardens for a farewell ceremony at which his successor, Dr G.P. Darnell-Smith, spoke of his achievements. Several speakers expressed their disappointment at the Minister's unwillingness to make the farewell presentation in the appropriate setting of the Gardens. Superintendent Edward Ward advised that a wallet would be presented to the former Director at a separate private function in the office of the Minister, Frank Chaffey. In any case, it was considered to 'be more of a family gathering if they had Mr Maiden to themselves'.[22]

English by birth and in manner, Maiden long declared himself an Australian, and reinforced this by devoting much of his professional life to the study of eucalypts, thereby beginning a tradition which is still maintained. He combined the best qualities of a devoted public servant with those of a devoted scientist. He was a sound administrator, with a concern for well-ordered records, and was a faithful custodian of what he saw as his public trust, being prepared to defend it, if necessary, against political threat. As a botanist, he earned wide recognition, being honoured by a Fellowship of the Linnean Society of London in 1889 and by its gold medal in 1915; by a Fellowship of the Royal Society of London in 1916, the Mueller Medal of the Australasian Association for the Advancement of Science (1922) and Clarke Medal of the Royal Society of NSW (1924). He claimed not to 'make many friends' although he had 'plenty of acquaintances',[23] finding recreation in his work, and to a lesser extent in philately, a logical result of his worldwide correspondence. Within his area of jurisdiction, he ran a tight ship and paid public tribute to the fact that its crew 'had never been accused of "loafing"'. A good judge of staff potential, he gathered round him a team of workers who shared his sense of dedication. At his farewell, he told the staff: 'I have never willingly humbugged a man in my life. I tried to do my best for you all, and I have never told you a lie.' He also invited them to visit him at

Farewell of J.H. Maiden, 2 October 1924.
Seated *(l. to r.)*: *J.H. Maiden and Dr G.P. Darnell-Smith.*
Standing *(l. to r.)*: *William Hardie (Superintendent, Centennial Park); John Morgan Nichol (Superintendent, Campbelltown Nursery); Edward Naunton Ward (Curator, Botanic Gardens); William Grant (Superintendent, Domain); Richard Nichol (Secretary).*
(by courtesy of the Director, Royal Botanic Gardens, Melbourne)

Turramurra. Every member of the staff filed by the small, seated figure to shake hands and wish him well.[24]

There was not much time to enjoy the new garden at Turramurra or to proceed with the *Critical Revision.* Joseph Henry Maiden died at his home on 16 November 1925 from heart disease and arterio-sclerosis. A former warden of St Andrew's Cathedral, he was buried in the churchyard of St John's, Gordon, following a service attended by a host of mourners representing the width of his departmental, political, scientific, historical and other associations.[25]

The old Gardens residence still stands, and it is not difficult to imagine the diminutive, dynamic Maiden in its well-stocked study, the floor strewn with shoe-boxes bulging with envelope files, working on the *Forest Flora,* the *Critical Revision* or on yet another scientific paper, while a privileged cat sat at his elbow, giving the creative pen an occasional nibble. It is a warm and pleasing picture to consider especially when strolling along the memorial Maiden Walk, or resting in the Maiden Memorial Pavilion, or waiting in the J.H. Maiden Theatre for an enlightening lecture to begin.[26] For Joseph Henry Maiden, Sydney's Botanic Gardens occupied 'classic ground'.[27] With equal conviction, he could have said 'hallowed'.

Chapter 7

OLD OBJECTIVES, NEW DIRECTIONS 1924–85

... although the Gardens are magnificent and a priceless part of our heritage, we were inclined to take them so much for granted ...

THE HON. NEVILLE WRAN, 1982 [1]

On 1 July 1924, three important appointments took effect. Edward Naunton Ward, Superintendent of the Gardens since September 1914, became Curator of the Gardens and Centennial Park, Edwin Cheel became Curator of the Herbarium, and George Percy Darnell-Smith, foundation head of the Biological Branch of the Department of Agriculture, assumed the additional duties of 'Director of the Botanic Gardens, and Officer-in-Charge of Centennial Park.'[2]

Dr Darnell-Smith, born in Oxfordshire in October 1868, graduated from the University of London in 1891 with honours in botany and zoology. In 1918 he was awarded a doctorate by the University of Sydney for work on the lifecycle of the clubmoss or fork-fern, *Psilotum*. His training, previous career and writings, especially in the *NSW Agricultural Gazette*, reflected the interests and abilities of a practical departmental biologist rather than those of a research botanist. As he told Maiden, he was more interested in 'ecology and physiology' than in 'purely systematic work'. Although Darnell-Smith shared Maiden's keen interest in chemistry and technical education, he neither demonstrated the same dynamic energy nor made a comparable impact.

Belatedly recognized in Australia for the successful control by copper carbonate of bunt disease in wheat, Darnell-Smith brought to the Gardens the experience of an academic, a microbiologist and a pathologist.[3] As the first graduate Director, he also brought a new regard for appropriate academic training and qualifications.

Much of Darnell-Smith's directorship, (1924–33), was overshadowed by the Great Depression, but some significant physical and administrative changes were effected.

One of his first concerns was the Domain, for the Under Secretary and Director of Agriculture, George Valder, suggested that the terms 'Inner' and 'Outer' be dropped. This was promptly referred to the new Curator, E.N. Ward, who pointed to the Inner Domain's long association with Government House, and to the problems of separate votes and allocation of staff. Having had 'considerable experience under no less than nine Governors', Ward felt that it 'would be a pity to disturb' a system which worked 'so smoothly and economically'. Darnell-Smith concurred, and Valder was advised accordingly.[4]

The Under Secretary was also concerned with smooth and economical procedures, and the new Director's formidable work-load 'was arranged in four sections to deal with horticulture, systematic botany, agrostology, and biology, with a clerical section to handle correspondence, accounts, regulations, and library'.[5]

The Governor's old 'bathing house', built in 1826 near the north-western point of Farm Cove, 'entirely of Stone found on the Spot', was demolished by June 1924, and the stone used for a new fernery, opened in December.[6] 'Extensive alterations' were made to parts of the Gardens during 1924–27, including the relocation of Maiden's prickly pears, the expansion of the succulent garden, rearrangement of the southern Palace Garden opposite the Mitchell Library, renovation of rockeries near Victoria Lodge and enrichment of the azalea gardens on the south side of Macquarie's wall to make the 'Spring walk' even more attractive. On the other hand, adoption of the forty-four hour week in 1925, 'made it necessary to install electric mowers, and to do away with a number of flower beds, grass borders and verges'.[7]

At the more scientific level, W.F. Blakely, with R.H. Cambage's assistance, continued work on Maiden's unpublished material for the *Critical Revision of the Genus Eucalyptus*, and the seventy-fifth and final part was published in 1931. A collection of over 700 Tasmanian specimens from Professor A.H.S. Lucas was incorporated into the herbarium which was progressively rearranged according to the system of Engler and Prantl. Routine duties were maintained in plant identification, plant toxicity, weed eradication and farm forestry.

As the Depression loomed in 1928 and 1929, there was scrutiny of the dozen buildings in the Gardens, Domains, Centennial Park and Campbelltown Nursery occupied by departmental employees. Surveys were made of the amenities, valuations and rentals of these premises, even to the point of revealing that George Frederick Hawkey, superintendent at Campbelltown, still used kerosene lamps and 'made his own sanitary arrangements'. The new Director was further concerned by the suggestion that the herbarium building should be insured for only £5500. He valued it at £7500, the specimens at another £4000, and the library of some 10 000 volumes at £3000, in addition to furniture and other effects.[8] There was 'a reorganisation of staff, involving a greater use of mechanical power' to save over £3000 per year in labour and maintenance.[9]

Accountants derived little joy from the knowledge that in 1928 about 188 000 living plants had been freely distributed from the Gardens and nursery, and that the nursery was now graced by a rebuilt propagating house and a soil sterilizing plant. In 1929, the free distributions comprised over 50 000 shrubs and trees from the Gardens, and 120 000 from the nursery. In August 'a system of charging for plants ... was introduced' and predictably, the demand dramatically decreased. On 1 September

1930 the Campbelltown Nursery ended its half century of service, the superintendent and the stock were transferred to the Gardens, and the land offered for sale three months later. Some trees still stand as fortuitous memorials.[10]

Probes prompted by the economic emergency long continued. In March 1931 George David Ross, who had replaced George Valder as Under Secretary in 1926, undertook a survey of departmental laboratories to ensure that there was no overlapping of functions. Darnell-Smith promptly answered the query by declaring that the Gardens had no laboratory.[11] The Public Service Board sought detailed information about Edwin Cheel's duties, and the Director obliged with another memorandum. He also submitted details of the routine duties, general herbarium work, research, educational and miscellaneous duties performed by the indoor staff, stating that the collection then contained 'approximately half a million specimens, all ... labelled and arranged in strict botanical sequence'.

Despite the use of electric and petrol motors, some horse-drawn mowers were still being used in the Spring of 1931, and questions were asked about the retention of horses on the establishment when two motor lorries had been supplied. Darnell-Smith patiently pointed out that the horses were used for mowing and the lorries for transporting materials, equipment and refuse. By July 1932 memoranda were flowing freely from the Premier, B.S.B. Stevens, and his new Budget Committee, urging government institutions to keep requests and expenditure to the 'absolute minimum'.[12] In September, Darnell-Smith furnished an account of the outdoor staff, indicating how foremen or head gardeners, gardeners special class, gardeners and labourers were profitably occupied, defining their classifications and duties.

It took a Melbourne visitor to suggest one minor item of expenditure. He told the Director of 'two pleasant afternoons' spent in the 'fine garden', but had 'yet to find where the men's lavatories are hidden'. He had given up the quest twice, much to the delight of 'the girls in the tea rooms who always know what is being sought'. He suggested that a sign be erected, and the Director agreed. However, when the Australian Automatic Weighing Machine Company offered to supply the Gardens with 1000 plant labels annually in return for permission to install three or four machines, the Director was reminded that this would be against the regulations.[13]

Early in 1931, Darnell-Smith and A.R. Penfold, Curator of the Technological Museum, examined the possibility of promoting, at that cricket-conscious time, the cultivation of the Cricket Bat or Huntingdon Willow, *Salix alba coerulea*, which Ward had sent from England during a visit in 1927. There was also promotion of Tung Oil production from *Vernicia (Aleurites) fordii*, and the Director took a special interest in the possible production of Neem Oil from the Margosa tree of Ceylon, *Azadirachta indica*, to provide a 'fly-fuge and healer of wounds' in stock.[14]

During 1932, Darnell-Smith successfully applied to the Unemployment Relief Council for two grants of £300. The large pond in the Lower Garden was cleaned, its banks strenghtened, and the Director urged that a new wall be built round it using stone which could be obtained from excavations at the Harbour Bridge site. Another project was the renovation of an area near the conservatorium and its 'restoration to the public' after its closure 'for nineteen years in connection with the city railway'. Relief work also led to improvements in Centennial Park.[15]

By early 1932, W.H. Ward found that the tents and makeshift shelters of 'several hundreds of campers' in the Outer Domain posed a problem. Fencing, seats, and tree branches were consumed for firewood and there were difficulties with sanitation and garbage disposal. The situation was relieved in August, when many unemployed were moved from the Domain to the old fish markets at Redfern.[16]

In May 1932, Darnell-Smith sought permission to remove a cubic yard of brick-work near the herbarium entrance bearing the trachyte foundation stone of the 'Museum of Botany and Horticulture' laid with such promise on 13 June 1916.

Darnell-Smith, who had resigned as biologist on 1 July 1927, retired from Direc-torship of the Gardens in October 1933. Remembered as 'a most likeable bachelor' and an entertaining, chain-smoking raconteur with a direct but friendly manner, Darnell-Smith's sense of humour and regard for language were reflected in his corre-spondence. In May 1931, for example, he requested the departmental legal officer to frame a regulation to prohibit parking in certain areas by all vehicles, whether cars or dog-carts, and another 'to prohibit politicians and others from standing on statues' to deliver orations.[17] When in July 1932, the old bogey of cut flowers was raised, he dealt with it in true Maiden style. Flowers were sought for the Cabinet Room and daily for the Premier's Room. Darnell-Smith did not actually refuse the flowers, but simply urged 'that this request shall be withdrawn'.[18]

A strong advocate of Australian plants for Australian gardens, Darnell-Smith confessed to a 'Tree Week meeting' in August 1932 that the White Fig, *Ficus virens* (*F. cunninghamii*) of the North Coast rainforests was, to him, 'as beautiful as an oak'. He was conscious of past trends:

Looking back at the history of the parks and gardens of Sydney, we see that they have had four epidemics of tree planting—the Moreton Bay fig epidemic, the peppertree, the camphor laurel, and the Canary Island palm epidemic. Only one of these trees, the Moreton Bay fig, is a native of Australia ...

and even this, although handsome, required 'ample room ... to grow' otherwise he would not recommend it.[19]

Dr Darnell-Smith remained in Sydney after his retirement, and died at Manly in April 1942, when the Gardens were facing a threat which promised to be even more damaging than the Depression.

Dr Darnell-Smith's retirement in 1933 provided an ideal opportunity for adopting further economy. The office of Director was abolished, and administration of the Gardens and its associated reserves was shared by the two curators, Edward Ward (Gardens) and Edwin Cheel (Herbarium). This regrettable bifurcation continued for twelve years, with Ward being succeeded by George Frederick Hawkey in 1934, and Cheel by Robert Henry Anderson in 1936.

Responding to a request from the US Library Committee in June 1934, Edwin Cheel completed a questionnaire which provides an interesting picture of resources and activities of the time. The total area of the reserves was given as 972 acres, containing eleven greenhouses, with a combined area of 950 square yards, three 'con-servatories or exhibition buildings', with an area of 800 square yards, in addition to the herbarium, offices, etc. There were 107 employees, including four 'professional and scientific staff', four administrative staff and two clerical staff. Income from lava-tory fees and Domain sports grounds was about £330 year, while £2650 year went to maintenance and £25 000 to salaries and wages. The library remained steady at about

10 000 volumes, and the herbarium held 'upwards of half a million' specimens. Cheel declared that some plant breeding and hybridization were carried out, although the emphasis was on taxonomic research. Lectures were given, university students were assisted, there were links with the Royal and Linnean Societies of NSW and occasional radio broadcasts were made. Cheel declared that botanical research was the institution's chief function, although educational, economic and aesthetic roles were recognized and encouraged,[20] although evidence of the success of these laudable activities is rather sparse.

With the assistance of the Unemployment Relief Council, 'outside' work continued in 1934 and 1935 on cleaning ponds, repairing roads and improving drainage, while inside, considerable attention was given to weed control, especially of skeleton weed and burr plants. During 1936 and 1937 it was noted that landholders were showing an 'increased interest' in planting trees for windbreaks, shade and fodder. In addition to answering enquiries about arboriculture, herbarium staff produced two significant books: R.H. Anderson's *The Trees of New South Wales*, published by the Department of Agriculture in 1932, priced at 5 shillings and W.F. Blakely's privately published *A Key to the Eucalypts* which sold for 10 shillings. Further interest was stimulated by the launching of a tree planting scheme in Queen's Park on 27 July 1936.

Both Ward and Hawkey were approached between 1933 and 1935 by the Returned Soldiers League about germinating seeds from Gallipoli and the planting of seedlings around the new War Memorial in Hyde Park. Among the plants raised were pines from Lone Pine and Pine Ridge, broom from Quinn's Post and bog myrtle from Shrapnel Valley.[21] At the end of 1934 Cheel was asked by the British Legion Poppy Factory about supplies of the large Maiden Hair Fern, *Adiantum formosum* for memorial wreaths.[22] Even before these delayed echoes of the Great War reached the Gardens, staff were warned by the Defence Department that care was to be exercised should 'foreign powers' seek specialized and unpublished information.[23] There was always the chance that botanical knowledge could lead to new developments in drugs, lubricants and explosives.

The pre-Second World War 'relief work years' had their lighter moments. There was unsolicited advice on 'magnetic planting' to avoid production of 'Crooked, Bendy Hide bound, Curly and Ringy Trees', and the offer of a model crocodile 8 feet long and 6 inches wide which the Director felt was 'rather out of place', although Ward had it laid at the foot of a statue in the lower pond, before a presumably astonished Venus. In August 1935, as President of the Wattle League, Edwin Cheel received a letter addressed 'Dear Enemy of the Waratah as the National Floral Emblem of the Commonwealth', attacking anyone, including Dr Darnell-Smith, who had any doubts about wattle causing hay fever and catarrh.[24] Cheel, like Maiden, was a 'wattle man' who believed that the national emblem should grow nation-wide.

By April 1936 complaints made at a Public Service Association Conference that the Gardens were deteriorating, were taken up by the Premier's Department. Hawkey was understandably piqued. He wished 'to know in what regard it is considered deterioration has taken place'. He claimed that the Gardens were as well maintained 'as they have been for many years'; but if better maintenance were desired, this could be achieved by the allocation of two more mowers, the 'payment for all overtime' and the provision of labour to repaint all the fences and seats, since the establishment had 'only one painter, who is also the labelwriter'.

In September 1934 W.W. Froggatt wrote to the *Sydney Morning Herald* about the condition of the old Wishing Tree, and Hawkey received a letter on the subject from Thistle Y. Harris. It was suggested that a young Norfolk Island pine be planted before children could 'no longer try their luck round the old tree'. The matter was taken up with 'some of our children's organizations' and Lady Gowrie planted the replacement on 27 July 1935 just north of the gateway through Macquarie's Wall.[25]

When Robert Henry Anderson succeeded Edwin Cheel as Botanist and Curator of the Herbarium on 4 June 1936, he had to his credit several papers on weed control and farm forestry, some systematic studies of the family Chenopodiaceae, his book on trees, a part-time lectureship in Forestry at Sydney University, and fifteen years of dedicated service to the Gardens. Born at Cooma in March 1899, the son of a Presbyterian minister and educated at Fort Street High School, Anderson was destined to become the first Australian-born Director. After briefly considering an army career, Anderson became 'a consistently good student' in Agricultural Science, showing a special propensity for botany, and taking additional courses which almost qualified him for a second degree. His subdued soldiering ambition was partly achieved when he joined the AIF early in 1918, only to be returned from Capetown on account of the Armistice.[26]

Like Darnell-Smith, Anderson wanted to see an increase in well qualified scientific staff, and he was anxious that the Gardens should regain the function and reputation of an efficient scientific institution. Consequently he worked uneasily, yet loyally, in the divided administration, looking to the day when control would be unified.

Relief workers were still employed in the grounds late in 1937 and early in 1938 to prepare for the special displays and re-enactments of the Sesquicentenary Celebrations. These attracted enormous crowds to the Gardens and Domains from their inauguration on 26 January 1938 until Anzac Day, when the Domain was thronged with 'a concourse of people' perhaps never before witnessed. On 3 February Lady Gowrie officiated at the opening of the Pioneers' Garden on the site of the Garden Palace dome.

As part of his strategy to emphasize the scientific role of the Gardens and to promote the importance of research undertaken, Anderson initiated an institutional journal, *Contributions from the NSW National Herbarium*. The first issue appeared in July 1939 and was 'well received'.

It was appreciated quite early in the Second World War that locally produced substitutes would have to replace certain imported materials. Grazing industries assumed additional significance, and research into crop production, pasture improvement, weed control and poisonous plants became crucially important. Such pastoral and agricultural challenges were very much to Anderson's taste, while he continued to promote his tree planting views through farm forestry. In 1939 he was reclassified as 'Chief Botanist'.

During the first year of the European war, relief labour was used in the Garden, Domains and Centennial Park, cleaning ponds and channels and repairing roads and footpaths. War-time emergencies in 1940–41 were amplified by drought, and the famous azaleas were kept alive only by pumping water from Botanic Gardens Creek. The war made its first physical impact when, in the inner Domain, an opening was made into the railway tunnel to serve 'as an air raid shelter if required'.

The entry of Japan into the war on 7 December 1941 posed a near and formidable threat. The Herbarium and its precious contents were far too close to the obvious target of the naval docks on Garden Island, that once picturesque and peaceful part of 'the Governor's Demesne'. Anderson took immediate and positive action. He contacted the Director of Agriculture, Dr Robert J. Noble, and on 16 December formally advised that the type specimens in the herbarium were irreplaceable and with about 500 very rare and valuable books they should be taken without delay to a safe storage place in the country. Several places had been considered, but the Glen Innes Experiment Farm seemed the most appropriate. E.B. Furby, manager of the Experiment Farm was advised on the 18th, and the first consignment of '90 herbarium boxes containing approximately 2,600 specimens' left Sydney by rail on Friday 19th. This was a remarkable effort, for the specimens had to be extricated from the general collection. Anderson promised other shipments comprising the remainder of the type specimens, and then 'a specimen from each Australian species represented in the Herbarium'. In addition, original manuscripts of research work would be typed in multiple copies to be lodged in different places. During the weekend following 19 December one of the botanists, Douglas Oakeley Cross, (later a Macquarie Street allergist) went to Glen Innes to inspect the storage facilities, while war bulletins described the alarming rapidity of the Japanese advance.[27]

The hot dry Summer of 1941, the continuing water restrictions, and the increasing shortage of labour due to enlistments and wartime man-power allocations, caused the loss of quite a number of trees and shrubs and the abandonment of planting annuals. Another entrance to the underground railway tunnel was excavated in front of the State Library, part of Centennial Park was resumed for military use and other parts were transformed by slit-trenches and air raid shelters. With remarkable restraint, it was described as 'unfortunate that the Metropolitan Water Sewerage and Drainage Board saw fit to erect a war emergency pumping station' in the Domain, and 'a very substantial building' at that.

On 1 September 1942 G.F. Hawkey was appointed air raid warden of the Botanic Gardens buildings. Anderson had 'not the slightest objection' to this, but felt that he might have been consulted. After all, he worked in the herbarium which despite the transfers to Glen Innes, still contained 'one of the most valuable of the State's collections'. This required 'some special care and attention' and already Anderson had 'made certain arrangements for dealing with incendiary bombs'. He believed it was his first duty to safeguard the collection even if this responsibility were not officially recognized and he sought to be made warden or deputy warden of the herbarium building.[28]

Three months later, the tempers of both curators were tested by 'a comprehensive display of military equipment' in the Gardens, and lawns and footpaths suffered. Meanwhile some oil storage tanks were erected in the Domain as another emergency measure.

Routine work, like the drought, continued during 1943. At the end of the year, however, Anderson was delighted that the work of one of his honorary curators, the

Reverend H.M.R. Rupp (1872–1956) appeared as the first contribution to the much-desired new Flora of the State. This was *The Orchids of New South Wales* published with the aid of Orchid Society members and supportive citizens, and the generous personal (and anonymous) assistance of Miss Joyce Vickery, MSc, Assistant Botanist since 31 August 1936. Mr Rupp, who as a young man had met Baron von Mueller, shortly presented his 1500 orchid specimens to the herbarium, where he worked frequently almost until his death. A donation of a far different kind was formally acknowledged in November 1943. This was described by Anderson as 'a massive replica of the Choragic Monument of Lysicrates' erected in Athens in 335 BC. The replica, by Walter McGill, was moved from the grounds of Sir James Martin's former home at Potts Point where it had stood since 1870 until the property was acquired for naval purposes.

When the Pacific War ended in August 1945, part of the Domain was occupied by huts for about 2000 British sailors, but some facilities commandeered in Centennial Park had been restored to the department and reclamation work had begun. A month after the Armistice, the old Wishing Tree, then in a sad state, was finally felled and much of the timber went to Concord Hospital where ex-servicemen used it for making trinket boxes and other mementos. Leo and Ida Buring offered to mark the site by an appropriate sculpture, and Dr A.J. Fleischmann's 'I Wish' was commissioned.

G.F. Hawkey retired in September 1945, and at last re-unification of the administration began, with R.H. Anderson as Chief Botanist and Curator, Botanic Gardens and National Herbarium as from 5 October. David Ross Frame (b. 1888) was appointed superintendent of the Gardens and officer-in-charge of Centennial Park. Anderson considered his major task was to ensure 'that the Botanic Gardens are so developed that they should regain their position as a scientific institution of worldwide reputation'.[29]

During the final months of the war and the immediate post-war years, both inside and outside staff were kept busy answering an increasing number of enquiries, and meeting the growing demand for sports fields, restoring buildings and other facilities after years of little or no maintenance, and obliterating the physical scars of wartime emergency measures. Anderson revised his *Trees of New South Wales* and bulletin, *Tree Planting on the Farm*, and encouraged systematic research on ferns, grasses, wattles and noxious plants, and the completion of Mrs Alma T. Lee's major work on the genus *Swainsona* which occupied a full issue of the *Contributions* in July 1948. By June 1946 Anderson was committed to meeting the need for a complete revision of the flora of New South Wales, a specimen registration system was adopted, and with Joyce Vickery's enthusiastic assistance, a start was made on overhauling the herbarium library, from which material was sent to Manila to replace losses sustained during the Japanese occupation. Peacetime conditions permitted the revival of a beneficial exchange programme during 1946 and 1947. At that time a young Science honours student—Lawrence Alexander Sidney Johnson—came to work on a revision of the genus *Casuarina*, the she-oaks. He joined the staff as Assistant Botanist in February 1948.

In the grounds, which had been deprived of fertilizers during the war, some improvements were effected by using manure from the State abattoirs and the Royal Easter Show, and a campaign was waged against the plant pests which had become a threat in the absence of appropriate sprays. New shelters were planned for visitors, long-needed repairs to roads and footpaths were undertaken, and through the generosity of Mrs D. Cohen, the fountain erected near the herbarium in 1888 to the memory of Lewis Wolfe Levy (1815–85), prominent businessman, philanthropist and parliamentarian, was put in order after a long period of disuse, while the Statues Committee made an appropriate survey of the Gardens and Centennial Park.

By 1950 the astonishing deficiency of there being one compound microscope available for the whole staff had been partly rectified by the supply of the second such instrument and 'two binocular lowpowered bisecting microscopes' which facilitated the identification of an increasing number of specimens sent by enquirers. Systematic work was gathering momentum for the proposed 'Flora Series' of the *Contributions*, and with the aid of the recently appointed seed and plant collector, Ernest Francis Constable, the exchange programme had been expanded. In 1949–51, Miss Mary Tindale worked at Kew as the first staff member to be appointed to the important position of Australian Liaison Officer. In this role she enabled 'various important research problems' to be resolved by reference to essential literature and type specimens.

In the grounds, the first frost for over thirty years damaged some tropical plants in July 1948, and another occurred the following June. Power sprays, motor mowers and motor scythes were being used effectively, while the landscape gardener, Lionel Harold Steenbohm (b. 1915) worked on a detailed plan of the Gardens to supersede the last plan made in 1921. George Chippendale, who had worked in the herbarium and museum since 1936, was, at Anderson's suggestion, busily compiling a catalogue of living plants, the first since 1895.

Distinguished visitors in 1950 included the Director of Kew, Sir Edward Salisbury, who considered the palm plantation 'one of the finest ... in the world'. However, in the herbarium, he echoed the concern of the visiting Swedish botanist, Dr Johan Mauritzon, who, at the end of 1936, had expressed alarm at 20 000 cardboard boxes containing half a million naphthalene-sprinkled specimens being stored in a building which was a potential fire trap.[30]

During the 1950s systematic work proceeded on ferns, grasses, wattles and eucalypts and several other groups, and manuscripts were prepared for the proposed new Flora. Records of newly discovered introduced plants and of new localities of indigenous species were published, and the usual examinations of stomach contents were conducted when stock poisoning by plants was suspected. During 1952 these were made on an average of twice a week. Specimens poured into the herbarium at the rate of 1000 to 3000 or so annually through exchanges, and between 1952 and 1956 the private herbarium of some 10 000 specimens collected by Dr F.A. Rodway of Nowra came as an additional welcome donation. The museum which George Chippendale had rearranged into some order closed about the time he enlisted in 1941, and a decade or so later efforts were made to renovate the displays in a more modern and attractive setting, but although many of the old specimens and curiosities displayed in Maiden's time are fortunately still preserved, the museum has not been revived in the form then conceived.

The Gardens and Domains attracted special attention during national and Royal occasions during the 1950s. In January 1951, the Jubilee of the Commonwealth was celebrated, and the Gardens were opened on the night of the 29th for a Venetian Carnival, but 'the crowds proved uncontrollable' and damage resulted.

An agriculturist by training, Anderson appreciated the special historical significance of the Gardens. On 14 October 1952 a small monument beside the creek was unveiled to commemorate that 'In these Gardens began the agriculture and horticulture of a continent ...'. Near the conservatorium overlooking the Nellie Stewart rosery another memorial was completed in the form of a seat to commemorate G.P. Darnell-Smith.

By 13 February 1954, the Gardens were at their best for the landing in eastern Farm Cove of Queen Elizabeth and Prince Philip—the first occasion on which a reigning monarch had stepped on Australian soil. Two years later, the site was marked by a commemorative wall unveiled by the Governor-General, Sir William Slim. The Royal visit provided a fillip to requests to get things done, and Anderson noted with pleasure that 'special attention was given to lawns, shrubberies and trees, roads and footpaths were repaired, gates, fences and seats were painted, water services renewed and other important work was carried out ...'.[31] In Centennial Park, 20 000 schoolchildren welcomed the Royal visitors.

Among innovations of the mid-fifties was the planting of three beds near the sea-wall with species known to resist the sea winds and salt spray which had so frustrated Charles Moore in his efforts to develop the area he had reclaimed. New plots of medicinal plants were also established, while further afield, as threats to the physical integrity of the Gardens and Domains mounted, Anderson and his senior staff visited areas near French's Forest, Hornsby, Narrabeen, Cabramatta, Prospect and elsewhere seeking sites for likely annexes to the Gardens. Within the herbarium, botanists at last were given better light through fluorescent installations, together with a specimen drier, an additional microscope, and, remarkably, one 35 mm camera.

The late 1950s brought upheaval and despair. In 1956 work began on a City Council car park in the Outer Domain, at an initial cost of forty-seven trees, most well-grown and some 'comparatively rare'. In the Inner Domain, work began on the Cahill Expressway, and many trees and shrubs were lost, and more alarmingly, there were plans to extend the highway through the Gardens and Domain.

Queen Elizabeth, the Queen Mother, visited in February 1958 during a drought emergency which meant negotiating with the Water Board for the legitimate if limited use of a few hoses. Another visitor was from Natal, Dr John Beard, who was escorted to the Blue Mountains to study wattles. Meanwhile, the decision was made to put the expressway through the Gardens and Domain, thereby destroying the essential unity they had demonstrated since the days of Governor Phillip. Excavations began in December 1958, causing in the first six months the virtual destruction of Fig Tree Avenue and an initial loss to the Gardens of twenty-four palms and a dozen other trees. Two brothers recently appointed to their new positions, Maurice Watson as landscape gardener and Warwick Watson as assistant superintendent, found an immediate challenge.

Administrative as well as physical threats were posed. Since a reorganization of the Department of Agriculture in November 1941, the Gardens had been attached to one of seven divisions, that of 'Science Services'. Early in August 1958, it was proposed to appoint a new Chief of Science Services with headquarters at Rydalmere.

Anderson saw in this the danger that the Gardens would slide even further down the scale of importance in the departmental structure. He urged that the department's Botanical Branch and the Gardens should be united fully and given individual status within the department. To make this clear, 'consideration might be given to the desirability of altering the designation of Chief Botanist & Curator to Director, Botanic Gardens & National Herbarium'. After all, similar institutions had Directors as did certain branches within the department itself.[32]

Following a meeting of the Trustees another recommendation was made on 24 October 1958 that the Gardens, in view of their long history and association with Royal visits since 1868, should be granted the Royal epithet, and this was done the following year.[33]

While bulldozers, drills and excavators thundered, the number of visitors declined, but on the positive side, work proceeded on the new Flora, with instalments being sent in October 1957, June and July 1958 to the department for transfer to the Government Printing Office. Regrettably they were delayed until 1961 when the first parts appeared with an Introduction by Anderson, now at last designated 'Director and Chief Botanist'.

Anderson's last years as Director, 1960–64, were overshadowed by the massive excavation works and by the belief that he had not been kept fully informed. In June 1961 he recorded:

Further trees and shrubs were destroyed or removed because of the Expressway, including two more Moreton Bay Figs from the southern end of Fig Tree Avenue. This once magnificent avenue has been so reduced that retention seems hardly worthwhile. If there had been in the first place full information as to the extent of destruction contemplated by the planning authorities, it would have been possible to offer other suggestions for the route of the Expressway. Authorities intruding into the Gardens and Domain minimise the extent of their proposed operations to a point where they become quite misleading.[34]

In a chapter which lacked his customary whimsical humour, Anderson took up the theme in his *ABC of the Royal Botanic Gardens*, begun before his retirement and published just after it. He warned of the desperate shortage of parklands and playing fields in Sydney, and of the prospect of the creation of an asphalt jungle. 'There is no saying how far the invader will go' declared the disillusioned Anderson. In the Domain, for instance, 'in 1963, a pleasant grassed bank adjacent to the appallingly ugly oil tanks was alienated for the building of an electricity sub-station ...' and there was already 'a dreary overflow of the concrete, bricks and mortar of the inner city into the quiet parklands of the Domain'. The invader, in whatever guise, 'armed with excuses and self-deluded justifications, but above all by the instinct to get something for nothing ... continuously seeks to divert gardens and parklands from their proper usage,' and the citizens of Sydney had to be on their guard. The Cahill Expressway was opened for traffic on 1 March 1962, but Anderson considered the restoration work had been disappointingly slow despite plans for reconstruction of several areas.

Anderson made himself readily available to journalists seeking interviews and comment on current issues, and he thereby did much to publicize the work and the needs of the Gardens. On 22 June 1960, for example, he was reported as having described the herbarium as a fire trap where a State collection of incalculable value would 'burn

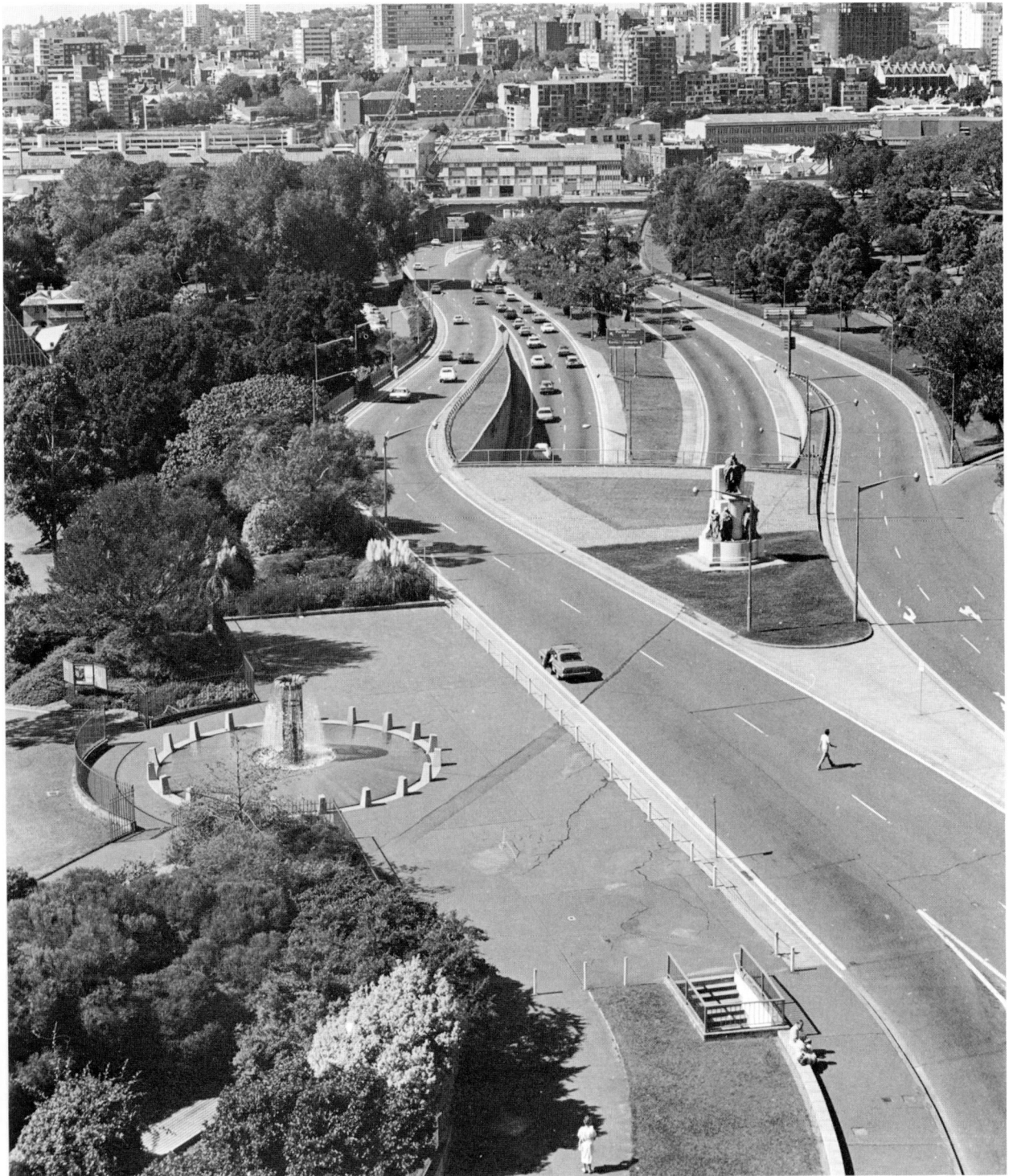

View in 1985 from the State Office Block in Macquarie Street showing the effect of the Cahill Expressway in destroying the unity of the Gardens and Domain.
(David Bedford)

furiously' in the event of fire.[35] The following day, the Minister for Agriculture, R.B. Nott, was reported as having sought £50 000 for 1960–61 to cover the cost of a new herbarium.

A little-known aspect of work in the herbarium received publicity when Dr Joyce Vickery, an expert on grasses, provided forensic evidence in the Graeme Thorne kidnapping and murder case. For this and other services she was awarded an MBE in 1962. From the early 1960s the identification of plant materials assumed a new and urgent importance during the upsurge of drug trafficking, and marijuana or Indian hemp plants, *Cannabis sativa*, were raised in the Gardens for exhibits in the Customs, Police and Health Departments.

Other developments during the early 1960s included the long-desired installation of an automatic sprinkler and fire alarm system in the herbarium. This relieved one anxiety, but did nothing to alleviate the serious problem of inadequate accommodation, either for staff or specimens. Nevertheless, research continued and additional parts of the Flora Series were prepared, the task aided by Mr L.A.S. Johnson's appointment as Liaison Officer at Kew for 1962–63.

In October 1962, Anderson and his scientific staff became involved in a work-value case before the Industrial Commission. They gave evidence in portentous company. One of the the judges, Mr Justice Alexander Beattie, was to become the first Chairman of the new Royal Botanic Gardens and Domain Trust; Mr Lionel Murphy, QC (later Federal Attorney-General and High Court judge) assisted by Mr Neville K. Wran (later Premier of New South Wales) appeared for the Public Service Association; Mr John R. Kerr, QC (later Governor-General) assisted by Mr K. Cohen appeared for the Public Service Board. In the course of the enquiry Mr Knowles Mair, Senior Botanist, spent seven hours giving evidence and Mr Wran was conducted through the herbarium to see the nature of the work being performed under cramped and ill-equipped conditions. It was a salutary introduction to the scientific aspect of the place he had first visited as a child on picnics in the 1930s. A new award was made for departmental scientific officers, and the commissioners especially commended the work of Dr Joyce Vickery and Dr Barbara Briggs.[36]

The observation made at the enquiry that 'four of the [five] full-time botanists are females' highlighted the fact that since the appointment of the unlucky Miss Hynes in 1900, several very accomplished women have pursued long and successful full- or part-time careers at the herbarium, more especially during the last half-century. Some of them gained that curious departmental classification of being 'employed though married' as if the state of matrimony were apt to terminate various forms of thought and activity previously enjoyed.

Dr Joyce Winifred Vickery (1908–79) whose work has been mentioned, was appointed Assistant Botanist in August 1936, 'the first woman ever appointed a professional officer in a scientific capacity in the Public Service'.[37] She retired in 1968 as a widely-acknowledged authority on grasses, and continued working in an honorary position for the remainder of her life.

Mrs Alma T. Lee (née Melvaine) first worked in the herbarium as a Science honours student in 1934. She returned in 1938 to work on fungi and ferns before joining the Council for Scientific and Industrial Research in Canberra. Subsequently Mrs Lee returned from 1941 until 1946 and again from May 1959 as part-time Assistant

Botanist until her retirement on 30 June 1981, when she became an Honorary Research Fellow. She is well-known for her monograph on the genus *Swainsona*, a taxonomic model for its time (1948), and for revisions of the genera *Lomandra* and *Xanthorrhoea*, work which has been continued by Mr David Bedford.

Mrs Valerie Jones (née May) began her long association with the herbarium in 1937, also as a Science honours student. While employed by the CSIR she worked part-time at the herbarium, specializing in algae, first seaweeds, then freshwater forms. Her research became especially important during the war when new sources of agar had to be found.[38] Mrs Jones has published many papers, under her maiden name, in the *Contributions*, *Telopea* and elsewhere, and as Honorary Phycologist (formerly Algologist), her knowledge and skills are frequently sought to identify and resolve problems such as water discoloration or toxicity arising from algal growth.

Dr Mary Tindale joined the staff in April 1944 and became an authority on ferns, 'fern allies' and wattles, with numerous publications and much editorial work to her credit by the time she retired in July 1983. She continues her investigations and writing as an Honorary Research Associate.

Mrs Joy Thompson (née Gardiner-Garden) was appointed late in 1946 primarily to work on 'The Flora of New South Wales' but in fact she became 'a general staff botanist'. Leaving the herbarium early in 1959, Mrs Thompson returned as a part-time botanist some ten years later. In 1956 her revision of the native Cypress Pines, genus *Callitris*, was published in the *Contributions*, and more of her work appeared in the first part of the Flora Series in 1961, followed by a major paper on the family Papilionaceae (1961) with the second part (compiled with Mrs A.T. Lee) issued in 1984 as Fabaceae. Mrs Thompson has also made revisions of the families Polygalaceae and Tremandraceae and a detailed botanico-ecological survey of the Kosciusko region.[39] She is currently revising the genus *Leptospermum* for the *Flora of Australia*.

Yet another of the women who have made such an outstanding contribution is Miss Neridah C. Ford who between July 1947 and May 1971, earned the reputation of an industrious and meticulous staff botanist.

A later arrival was Dr Barbara G. Briggs who, having enjoyed gratuitous work in the herbarium during a university vacation, joined the staff in November 1959, filling the vacancy left by Mrs Thompson. Dr Briggs's special interest in cytotaxonomy is reflected in her work on various taxa of monocotyledons and on the families Plantaginaceae and Oleaceae, published in the *Contributions* and *Telopea*. The same interest has led to studies undertaken with Dr Lawrie Johnson, of the families Proteaceae, Myrtaceae and Restionaceae. She has served as a member and chairman of the editorial committee of the *Flora of Australia*.

Robert Anderson retired on 12 March 1964, his sixty-fifth birthday, after forty-three years of service. A self-styled 'underground engineer' whom his colleagues recognized as a shrewd negotiator, he left the reputation of an approachable administrator who did not like ringing bells for people, and who despite troublous times, maintained high staff morale, through encouragement, advice, the provision of opportunity and even financial assistance. Totally committed to this work and blessed with 'a confessed and impish delight in cutting red tape',[40] Anderson declared: 'If I had my time over again, I would do just the same—I could ask for nothing better'.[41]

As a perceptive reporter noted, the affable tree-loving Director, with his neat attire, and 'iron grey hair en brosse', did not conform to the popular conception of a public servant. He did not even speak in Public Service jargon and was 'old fashioned enough to believe that the job of a public servant is to give service to the public'.[42] When Anderson died at his Chatswood home on 17 August 1969, his long-standing colleague, Dr Joyce Vickery, lamented that botanists and horticulturists, both in the Gardens and beyond, had 'lost a warm-hearted, kindly and lovable friend'.[43]

Anderson was succeeded by Herbert Knowles Charles Mair, who was born at Toongabbie in 1909 and graduated from Sydney University in 1932 with first class honours in Botany. Thereafter he served as agrostologist in the Division of Plant Industry, Council for Scientific and Industrial Research in Canberra; Superintendent of Agriculture for the Northern Territory and Curator of Darwin Botanic Gardens; Commonwealth Research Fellow, investigating flax, at Sydney University; and as a staff officer in the AIF responsible for supervising army farms' production of vegetables and fruit in the Northern Territory and New Guinea; and later, as forest botanist in New Guinea and Borneo. He joined the Gardens as Assistant Botanist in May 1946, and in December 1948 became Senior Botanist, to which position Dr Joyce Vickery succeeded him.

A friendly man with an engaging manner, Knowles Mair moved easily into the new position, being well-known to the staff and an admirer of Anderson and his policies. One of the first tasks was to expedite the restoration, or at least the improvement, of grounds devastated by the expressway construction and the Domain parking station. Large areas, previously denuded of turf by shade, erosion and concentrated traffic were restored, and attention given to top-dressing (using the time-honoured source of silt from the Lower Garden ponds), and to the improvement of drainage and the extension of native plantings.

Mair was a foundation member of the Centennial Park Supervisory Committee, established in May 1964 on the initiative of Mr Harry Heath of the Public Service Board. During its four years of existence, the Committee obtained substantial grants for wide-ranging developments in both Centennial Park and the adjoining Queen's Park. In April 1965 Mair welcomed the Director of Kew, Sir George Taylor, who like his predecessor, Sir Edward Salisbury, greatly admired the palm grove. A gratifying task was to finalize acceptance of land at Mount Tomah for a satellite garden.

The post-war research activities which became more evident once Anderson had contrived to establish the herbarium's own publication, were vigorously maintained under Mair's encouragement. An increasing number of research papers for the *Contributions* and revisions of plant families for the Flora Series provided a constant challenge to the Government Printing Office. Lists of newly or rarely collected species were published, extensions of distributions recorded, and notice made of more sophisticated taxonomic methods such as cytotaxonomy and chemotaxonomy. Long-term studies of *Eucalyptus* and *Acacia* were developed respectively by L.A.S. Johnson and Mary D. Tindale. In 1966 and 1967, information on plant toxicity was reorganized and mimeographed and each botanist took home a copy in case of emergency calls involving suspected plant poisoning especially among children.

In the old herbarium, working conditions deteriorated as the number of specimen boxes increased.
(Royal Botanic Gardens Library)

Such enthusiastic research, supported by the work of the botanical collector and exchange material from worldwide contacts, continued to emphasize the urgent need for better working and storage facilities. Mair's first report noted that 'much of senior officers' time was occupied by preliminary planning of a new Herbarium building, to house expanding collections and staff over the next fifty years'.

R.H. Anderson had hoped to expand the old herbarium by adding a wing or storey. Now the recommendation was for an entirely new building, a move supported by envious observations made at openings of new herbaria in Adelaide, Brisbane and Lae. Yet despite the production of Dr Joyce Vickery's models, the preparation of sketch plans during 1966 and 1967 and the submission of appropriately phrased requisitions, funds were not forthcoming, and disenchantment grew as successive departmental budgets were announced.

Problems also arose from discontent over pay, conditions and prospects, especially among the grounds and gardens staff. During the late 1960s, there was an alarming spate of resignations, rising to eighty-five in 1969–70, rendering continuity of some projects virtually impossible. This trend continued, with 100 resignations lodged during the year before the Public Service Board stemmed the flow somewhat by signing a new agreement in September 1974.

On the other hand, there were some whose long service was commended. For example, Charles Robert Gordon Grant who had joined the department in 1925, retired on his sixty-fifth birthday in June 1966, having served as Gardens Superintendent for his last fifteen years. He was succeeded by Warwick Rex Watson who had joined the staff in April 1951. Similarly, when the Botanical Collector, Ernest Francis Henry Constable retired in June 1968, he had completed twenty-two years of invaluable fieldwork, obtaining material for the herbarium, for exchange, for propagation, and for meeting the growing research needs of the scientific staff. He was succeeded by Robert G. Coveny, from the horticultural staff, who at the time of writing has already served seventeen years in the position.

While Mair appreciated Anderson's achievement in reuniting responsibilities for the herbarium and the Gardens within the Directorship, there remained administrative difficulties which he determined to resolve. In March 1968 the herbarium 'ceased to be the Botany Branch of the Division of Science Services' of the Department of Agriculture. It was now administratively combined with the Gardens to form an autonomous unit with its Director responsible solely to the Director of Agriculture as the head of a division in his own right. This was an important step towards the attainment of an autonomy which would lead to a new recognition of the institution and of its essential requirements.

Year by year, Mair pointed to the urgent need for a new herbarium, even taking the step in May 1969 of moving his office and those of senior horticultural and clerical staff out of the old building into the former Gardens Residence which had been renovated, thereby temporarily relieving the serious overcrowding.

The Cook Bicentenary offered a bright prospect for forcing the issue, and recommendations were made that the Queen should lay the foundation stone for a new building to mark the 200th anniversary of the visit of Banks and Solander. This proved to be a worthy, but vain endeavour. During a visit to the Gardens in September 1968, the Minister for Agriculture, G.R. Crawford, revealed that efforts to obtain a grant from Federal authorities had failed, despite the fact that 'an Opera House architect, Mr Peter Hall, had planned the new herbarium four years ago'. Instead, scarce loan funds went to Art Gallery extensions.[44]

Despite this mounting frustration, research work proceeded apace, the scientific staff enjoyed increasing academic and professional recognition and the herbarium continued to attract notable scientists from throughout the world. In January 1969, Mair took the initiative to appoint John Pickard as the first plant ecologist, who inaugurated ' a new line of investigation ... complementary to the long established taxonomic research'. Donald J. McGillivray, who had joined the staff in October 1964, was appointed Liaison Officer at Kew for 1969–70, and in June 1969 Obed David Evans (1889–1975) retired from the position of part-time botanist he had held since April 1959, only to continue for two more years to complete work on certain monocotyledonous families for the Flora Series.[45]

In the grounds, the turf demonstration plantings begun in 1963 continued to attract interest as new species were added; some damage to statues and fountains caused by acts of vandalism was made good; the Henry Kendall Memorial Seat erected in 1940 under a bequest from Mrs A.M. Hamilton-Grey was relocated beneath a fine hybrid plane tree in the Upper Garden; improvements were effected to stonework in the Pioneers' Memorial Garden; and from November 1966 visitors entering by the gate opposite the State Library enjoyed the sight of the Sir Leslie Morshead Memorial Fountain. The Outer Domain continued to provide parking for Christmas shoppers, emergency parking for desperate commuters during transport strikes, fields for some 1500 organized annual league, soccer and basketball matches, and space for the assembly and dispersal of processions and parades, as on Anzac Day. The damage to turfing was constant.

Mair became increasingly concerned about the deterioration of several of the old glasshouses. They were essential facilities, but fragile, and the state of some posed aesthetic and economic questions about their retention. It was resolved in 1969 that they should be replaced by 'a complex of modern environment-controlled units in an area more prominent and accessible to the public'. Mair discussed the matter with Warwick Watson, and by June 1970, Public Works architects had 'produced a plan for a new pyramid-shaped glasshouse' intended to be the first of a series providing a range of micro-climates. Constructed during 1970–71 by E.A. Watts Pty Ltd, of Sydney, at a cost of $120 000, this first pyramidal glasshouse in the world immediately became 'a focal point of visitor interest'.[46]

Successive Directors, Robert Henry Anderson (seated) (1945–64), and Knowles Mair (1964–70).
(by courtesy of Land *newspaper)*

159

The Gardens and Domains, and more especially the Banks-Solander specimens, assumed due prominence during the Cook Bicentenary celebrations in April 1970. Part of the Gardens was opened for the harbour carnival on the night of 29th, but while every precaution was taken, damage to plants inevitably occurred.

Having published an outline history of the Gardens[47] and remained at his post to see the fruition of many deliberations on, and preparations for, the Cook Bicentenary, the genial Knowles Mair now felt it was a fitting time to end his relatively short, but very significant directorship.

Knowles Mair retired at the end of June 1970, after twenty-four years at the Gardens, the last six as Director. The Deputy Chief Botanist, L.A.S. Johnson served as Acting Director until John Stanley Beard took up duty as Director on 6 October 1970.

Born at Gerrards Cross, Buckinghamshire in February 1916, Dr Beard was an Oxonian with a degree in Forestry, and higher degrees in ecology and vegetation mapping. Courteous and personable, Beard had worked in forestry in the West Indian region before serving as Estates Research Officer with the Natal Tanning and Extract Company, 1947–61. Coming to Australia in 1961, he was foundation Director of King's Park and Botanic Garden in Perth for nine years,[48] and became a strong advocate of a new *Flora of Australia*.[49]

Whereas in Perth he had contrived to establish a new project with hand-picked staff, in Sydney Beard found a long-established institution attached to a government department with set rules regarding employment and rather inflexible statements of duties. Beard found this hard to accept, for he was an independent soul with firm views and grand visions. He hoped to open parts of the Gardens to present sweeping vistas; Botanic Gardens Creek, that 'polluted drain', would be put underground and the creek bed converted into an oranmental water garden; a 'magnificent fountain' would play in Farm Cove, around which the walkway along Moore's wall would be lit and opened at night while music from a pontoon orchestral stage would delight the walkers and those dining in a terrace restaurant high above the wall commemorating the Queen's landing. He hoped to withdraw Maiden's old residence from administrative use and restore it to its original purpose. He wanted to encourage the Japan-Australia Association to fund a Japanese Garden—after all, one had been established in 1917.

Beard found that the herbarium was functioning well despite 'the difficulties under which the botanists worked, the whole place stacked to the roof with specimens, little working spaces found in front of each window, and the fire hazard indescribable'.[50] He was delighted with the number and qualifications of the herbarium staff, but officially declared: 'the building erected in 1899 to house one botanist and a few thousand specimens, today houses over a dozen botanists and nearly a million specimens'.[51] Agitation for a new herbarium was revived in a manner not entirely within the department's experience of dealing with the same embarrassingly frequent request.

Fortunately there were brighter aspects. Ronald Grewcoe was appointed first supervisor of the Mount Tomah Garden in April 1971; Mr L.A.S. Johnson was awarded a doctorate and his work with Professor L.D. Pryor on eucalypt classification was published after a decade of preparation; a committee which included Dr Beard and Professor Michael G. Pitman investigated the deaths of Norfolk Island pines at Bondi and Manly and traced the cause to detergents being discharged into the sea, only to be blown back as foam on to the trees. Sir Garfield Barwick, Mrs Margaret Davis, Mrs Elizabeth Price and others were happy to support the Director's suggestion that a group of 'Friends' be formed, but the time was not deemed appropriate. The way was prepared for the appointment of an Education Officer to deal with the increasing number of school visits, and the conversion of some beds near the herbarium to serve as teaching plots was mooted.[52] Warwick Watson, Superintendent of the Gardens since June 1966, was awarded a Public Service Board Fellowship to study parks and gardens in Europe and North America, March-July 1971, and he presented a voluminous report with recommendations on such matters as the improvement of garden displays, development of educational services, establishment of an arboretum and more effective horticultural training.[53]

Although research and publication, advisory services and ground improvements were all well maintained, Dr Beard became increasingly disillusioned about the gap between what was procedurally and departmentally acceptable and what he believed was culturally and scientifically desirable. Instead of confirming his appointment in the usual way, the department offered the Director a senior research post at the conclusion of an extended probationary period. Thereupon Dr Beard resigned as from 16 June 1972, and returned to Western Australia to resume work on the ecology and vegetation mapping of that vast State. Perhaps he came to Sydney a little too early.

Dr Beard was succeeded by his Deputy Chief Botanist, Dr Lawrie Johnson, who took office on 29 June 1972, three days after his forty-seventh birthday. Born at Cheltenham (Sydney) and educated at Parramatta High School and Sydney University, Dr Johnson had been associated with the herbarium for twenty-five yers. He understood the benefits and the difficulties of administering the Gardens as a division of a State department; he knew the special interests and abilities of his colleagues, and appreciated the potential of the Gardens as a scientific institution worthy of world acclaim. Johnson shared the interest of his predecessor in plant distribution and ecology, the educational potential of the Gardens and the proposal to publish an entirely new *Flora of Australia* to succeed Bentham's *Flora Australiensis*. He had close associations with the Linnean Society of NSW, the NSW Government's Parks and Reserves Scientific Committee, and other scientific and advisory bodies, and his approach was largely determined by the desire for better integration of the scientific and outdoor activities to enhance the botanical and educational emphasis in the collections of living plants.

From the outset, the new Director wanted the functions of the institution to be clearly understood. The Gardens and herbarium provided 'a broad coverage of advisory information and research services in botany and ornamental horticulture'; the satellite garden at Mount Tomah enabled the growing of cool-temperate plants, and the Domains and Centennial park served as 'large parks catering for passive recreation, sporting activities, and wildlife conservation'. Research activities were directed towards systematic botany, ecology, and 'the production of a detailed and comprehensive *Flora of New South Wales*'. This message was constantly repeated.[54]

One of Dr Johnson's immediate concerns was, of course, to continue agitating for a

Dr Lawrie Johnson, Director (1972–85).
Norman Carter's portrait of J.H. Maiden hangs above the fireplace.
(Rowan Fotheringham)

new herbarium building where much of the research function could be creditably performed. In December 1971, the fungus collections had been transferred on permanent loan to the Biology Branch at Rydalmere. This left a small area to be utilized when the remaining collections in over 20 000 boxes were completely rearranged. The most frequently consulted (Australian) material was made more accessible, but the overall problem of gross overcrowding remained, despite the further transfer of five staff and sections of the collection to Gardens Residence No.4, and the 'seed stocks for exchange ... to a former air raid shelter'. The Director gibingly declared that such makeshift arrangements for long-suffering staff and three-quarters of a million specimens were in stark contrast to the situation in 'other States, where major herbaria have been provided with modern buildings'. Sydney's old herbarium, containing a priceless national asset, was almost at 'the limit of its capacity'. The seriousness of the situation was reiterated with firm if tedious regularity from 1972 to 1975.

Nor was the momentum of other projects allowed to decrease. Investigation of the loss of Norfolk Island pines on metropolitan beaches was maintained by a Premier's committee which included the Director, the Superintendent (Warwick Watson) and the Ecologist (John Pickard), and a report was presented. The case for an education officer was pursued, and Mrs Betty J. Jacobs was seconded from the Department of Education. Following a feasibility study in 1972–73, a demountable classroom was erected near the herbarium and in the first year of operations, nearly 7000 school pupils in 175 groups were introduced to this new educational service. Mrs Jacobs was succeeded in turn by Mrs Diane Alford in January 1977 and Mr John Johnstone in September 1980. Educational plantings were established in the Middle Garden, the site of the old Government Farm, to illustrate culinary uses, morphology, adaptation and relationships of plants in informative 'Plants, Evolution and Man' displays. Thus were vegetables overtly grown in the area for the first time since Allan Cunningham!

During the 1970s the problem of retaining Gardens staff continued, although in January 1975 the first eight horticultural apprentices began duty, including 'the first female outdoor worker', Ruth Murray, who qualified as a gardener four years later. By June 1977, thirteen Hong Kong fellows had undertaken intensive training as part of an overseas assistance plan.

There arose more frequent opportunities for scientific and horticultural staff to attend meetings and conferences from which much could be learned, and at which the Sydney Gardens could be promoted. The Director and senior staff members represented the Gardens at gatherings throughout Australia, in Europe, Asia and America, on both official and private journeys. In August 1975, Donald F. Blaxell became the fourth staff member to be appointed Australian Liaison Officer at Kew, and he joined the Director and Deputy Chief Botanist, Dr Barbara Briggs, at the XII International Botanical Congress in Leningrad, where Australia's invitation to hold the next congress in Sydney in 1981 was accepted.

Through negotiation with the Public Service Board, two senior positions were re-classified, and in July 1976 Dr Briggs became Assistant Director (National Herbarium) and Mr W.R. Watson Assistant Director (Gardens). This was a significant step forward in restructuring the administration to meet growing demands. Even the problem of overcrowding in the herbarium building showed promise of resolution at last when after prolonged discussions with the special projects architects of the Public Works Department during 1976 and 1977, 'detailed sketch plans of the proposed new Herbarium' were prepared.

By June 1978, further research by various staff botanists on the Myrtaceae, Proteaceae and other major families had been completed, some to the publication stage; work on wattles, ferns and algae was maintained, and a vegetation map of Lord Howe Island had been completed, while the mapping of the Cumberland Plain and north-west of the State proceeded. The publication programme was also revised. The *Contributions*, first issued in July 1939 with Miss Joyce Vickery as editor, ceased publication in December 1973, when Dr Mary Tindale transferred her editorial skills to the retitled journal *Telopea*, which began in 1975. The Flora Series of the *Contributions* instituted in 1961, and later renamed *Flora of New South Wales*, was maintained until 1984, when it closed with the paper on the legume Family Fabaceae. A new serial, *Cunninghamia*, devoted to ecological issues including vegetation mapping, was launched in 1981 with Dr Jocelyn M. Powell as editor. The second issue, 1983, was devoted entirely to the meticulous work of John Pickard and Anthony Rodd on the flora and ecology of Lord Howe Island. In 1981, the herbarium published *Plants of New South Wales: A Census of the Cycads, Conifers and Angiosperms* by Dr Surrey W.L.

Jacobs and John Pickard, the first such work since Maiden and Betche's *Census* of 1916, and a basic reference for students of botany. Currently, the contributions of several botanists are being incorporated into the new *Flora of Australia* being compiled in Canberra and the new identification manual of NSW plants is being prepared at the herbarium under the editorship of Mrs Gwen Harden.

The late 1970s were marked by political and administrative developments destined to have a profound and enduring effect on the Gardens and ancillary reserves. On 1 May 1976 the Honourable Neville K. Wran, QC, who had inspected working conditions in the herbarium for the industrial proceedings of fourteen years before, became Premier of New South Wales. It so happens that the view from the Premier's office in the State Office Block embraces a magnificent view of Farm Cove, the Gardens and Domains. As Premier, Mr Wran often visited Government House and other properties which had long associations with the Gardens. As a citizen, he lived close to Centennial Park. He viewed these reserves as valuable public resources, and amid the constant agitation for better facilities and the difficulties in maintaining a full work-force, he felt that the institution which was literally and constantly under his eye 'needed livening up'. The image of the Gardens seemed to reflect the archaism of some of the existing regulations, perhaps attributable to the attachment to a State department which had developed a vast range of responsibilities.

Accordingly, in October 1978, the Royal Botanic Gardens organization was transferred from the ministerial responsibility of the Minister for Agriculture to that of the Premier. After seventy years, the nexus between the Gardens and the Department of Agriculture was broken, an action welcomed by the Director, who had in fact broached the matter with the Premier.

On 31 January 1979, during discussions with the Premier, the Director offered to report on what he considered were the aims, functions, resources and needs of the institution. This provided a timely opportunity to outline the administrative structure, the scientific, horticultural and educational functions, the services provided to government, industry, similar institutions and an enquiring public, the current deficiencies and likely remedies. The principal needs were declared to be a new herbarium building, additional staff, acquisition of land in the Campbelltown area for an arboretum, and funds to develop the garden at Mount Tomah.

The Gardens were transferred administratively to the Premier's Department on 30 March 1979, and work began on a Bill for a Royal Botanic Gardens and Domain Trust Act. Introduced by the Premier, the Bill was debated in February 1980. The measure was supported by the Opposition, and some members on both sides indicated that they had done appropriate homework. The House resounded with the names of Cook, Banks, Phillip, Macquarie, Fraser and Cunningham, and there were odd gems of repartee. When mention of Sir Joseph Dalton Hooker prompted the question of whether he were an ancestor of a well-known real estate family, the Premier wryly observed: 'He could not be, otherwise the land would have been subdivided'.

The quip was not without its point, for one of the stated purposes of the Bill was to prevent further encroachments on the dedicated land, which was declared to have already been reduced from 72.06 hectares in 1916 to 68.97 in 1955 and 63.14 in 1980. Sydney was declared to lag behind comparable cities in its area of parklands, and the

Cahill Expressway, the Domain parking station and the Commonwealth oil storage tanks all received restrained criticism. The hope was even expressed that 'a large part of the Cahill Expressway ... exposed to the sky' should be covered to enable the establishment of lawns.[55]

The Act received Royal Assent on 15 April 1980,[56] and the five members of the new Trust under the chairmanship of Sir Alexander Beattie (one of the former trustees under the Crown Lands Consolidation Act) first met on 1 July,[57] taking over from the previous Trust, of which the Secretary of the Premier's Department, Mr Gerald Gleeson, had been the last chairman.

Thenceforth, the *Annual Reports* on the Gardens and Domains even surpassed in information and illustration those in which J.H. Maiden took such care and pride. Future historians will have ample reason to be grateful, but they no longer refer to Centennial Park, for, with the Director's encouragement and blessing, it had been dissociated from the Gardens after some ninety-two years to be controlled by its own authority by June 1980.

There were immediate and momentous developments, some proceeding from earlier initiatives and some from entirely new ones. Early in 1980 work began at last on a new herbarium building to cost in excess of $4 000 000 and to be named in honour of Robert Brown, the botanist who visited the colony with Lieutenant Matthew Flinders on the *Investigator* in 1802. Progress was made at Mount Tomah, and the new Trust established one committee to review scientific work and policy and another to report on the library. An important administrative reorganization, reflecting the Director's desire to enrich and thematize the Gardens plantings, established the new position of Assistant Director (Living Collections and Communication), to which Donald F. Blaxell was appointed in September 1981. Mr Blaxell's special interests include the eucalypts and orchids.

During April 1981, Charles, Prince of Wales, visited the Gardens and augmented the collection of 'royal trees' and the Premier launched the 'Great Pyramid Appeal' to raise funds for another glasshouse. Furthering its plan to foster and improve communications between the Gardens and the community, the Trust commissioned a new guide and information book, the first since Maiden's publication of 1903. Capably edited by Edwin Wilson, Extension Officer since August 1980, this well-produced work has enlightened many visitors since 1982. New signs were erected at the main entrances to attract, welcome and inform people rather than to present them with a list of caveats.

During the early 1980s the Gardens enjoyed further energetic promotion both locally and overseas, through press, radio and television, involvement in Heritage Week and in exhibitions in the State Library and Australian Museum, and through close association with the planning and staging of the XIII International Botanical Congress which attracted 2500 botanists to Sydney University in August 1981. In February 1982 the library and the botanical collections were moved into the recently completed Robert Brown Building. About a million specimens were transferred from the old cardboard boxes to new plastic containers, rearranged according to the most modern systems available, and the opportunity was taken to incorporate previously donated material from the Museum of Applied Arts and Sciences, Hawkesbury Agricultural College and Mr E.J. McBarron. The priceless botanical and literary resources of the institution were at last housed under conditions appropriate to their worth. Work proceeded on renovating the Anderson Building (the old herbarium) and the Cunningham Building (the former residence).

Staff morale and the new building rose together. Research and field work were reflected in the growing stream of publications. Vegetation mapping continued, and the ecology section was inundated by requests relating to local surveys, regeneration schemes and environmental impact studies, some associated with controversial highway constructions or logging operations.

The Gardens alone were now believed to be attracting about two million visitors annually, including nature and garden lovers, wedding parties, jogging enthusiasts, visitors to Festival of Sydney and Carnivale presentations, and to the 'Arts in the Gardens' Sunday displays along the Macquarie Street fence. Over two dozen horticultural and other apprentices were being trained, and scores of school students seeking work experience were being accommodated.

Sir Alexander Beattle retired as Trust Chairman in January 1982, and the Premier nominated Professor Michael Pitman, OBE, ScD, FAA, to succeed him. On 31 March the new Chairman welcomed HM King Carl Gustaf XVI of Sweden who dedicated the Solander Memorial Garden near the Anderson Building. Staff changes early in 1982 included the resignation of the knowledgeable Tony Rodd after some ten extremely productive years as Horticultural Botanist and his replacement by Dr Ben Wallace. A year later, Kenneth James Rendall, foreman, retired after forty-three years' service to the Gardens.

The Trust revived the idea of establishing a group of 'Friends' to support the Gardens. The first general meeting of some people 'with diverse but relevant interests' was held under the chairmanship of Sir Rupert Myers on 27 July 1982. The movement was immediately successful. A programme of activities was devised, and by the end of June 1983, there were 585 members. A year later there were over 1300.

The highlight of 1982 was the opening of the new herbarium building complex by the Premier on Saturday, 6 November, when it was noted that the file on a new herbarium dated from 1937. Due acknowledgement was made to Andrew Andersons, Brian Zulaikha, Michael Fletcher and colleagues in the Government Architect's Branch for producing, after consultations with scientific staff, a handsome and functional design for 'an eminently workable building and complex'. Special gratitude was expressed to the Premier, who recalled the tingling down the spine he had experienced at his first sight of the Banks-Solander specimens during an earlier visit. The government, he said, imposed only one condition—part of that historic collection would always be on display so that others might experience a similarly thrilling encounter with the past. The next day the Visitor Centre in the former museum room in the Anderson Building, dating from Moore's time and now tastefully refurbished, was opened to the public. With its bookshop and changing exhibitions, it has been a dynamic place ever since.

The new complex provided not only the realistic housing of the library, herbarium collections and staff, but also laboratories, darkrooms, a microcomputer, a scanning electron microscope, and special services for the public including the Public Reference Collection. Established in 1982–83, this collection comprises albums of specimens of over 1500 species, available to enquirers wishing to identify their own discoveries. This innovation is better for the enquirer, it furthers the cause of science and relieves the staff. Notwithstanding such revolutionary changes, shades of old problems linger—both Moore and Maiden would be intrigued to know that their old enemy, Nutgrass, *Cyperus rotundus*, still receives attention from their sophisticated successors!

In April 1983 the Trust was increased to seven members, and in July 1984 Professor Pitman was succeeded as Chairman by Mr John Edward Ferris, AM, BE, FIEA. Trust members, the Friends and the Volunteer Guides have worked enthusiastically and effectively to promote the Gardens into a new era of growth and esteem, thereby complementing the efforts of the staff. The 'Kew tradition' was ably maintained by Dr Surrey W.L. Jacobs as Liaison Officer for 1983–84 and in 1984 Miss Leonie M. Kemp assisted the Education Officer to make productive contacts with isolated schools in the far west of the State. Thus the institution develops and extension work proceeds, while the principal approach to the Gardens from Macquarie Street, with buildings dating from the time of Macquarie's Wall, is being transformed into one of the most elegant and informative historic thoroughfares in Sydney.

The full significance of Dr Lawrie Johnson's directorship, 1972–85, and of the individual labours of his large and competent staff, will be more properly appreciated and assessed at a future time in a wider historical perspective, but clearly it will be judged to have been of signal importance.

Chapter 8

BEYOND THE
GARDEN WALL

*The next generation will probably establish a country branch of the Sydney Botanic Gardens,
consisting of a readily accessible site of several hundreds of acres, within say thirty or forty miles of
Sydney. It will be away from the smoke of a large city, and it should contrast, as regards its soil,
with the natural barrenness of the Sydney garden.*

JOSEPH HENRY MAIDEN, 1912[1]

The first government gardens near Sydney and Farm Coves, while conveniently situated for the import and export of plants, were sadly deficient in deep fertile soil. When hopes for success in agricultural projects turned from Sydney to Rose Hill (Parramatta), the Governor's country residence was established there too, on the Crescent above the natural Amphitheatre, a well-known feature of Parramatta Park. In Phillip's time, this residence soon had its Governor's Garden which by October 1792 comprised $6\frac{1}{2}$ acres, including a 3-acre vineyard.[2] Early in May 1800 Governor-elect Philip Gidley King advised Sir Joseph Banks that he had also 'marked out a botanic garden, to be under Col. Paterson's directions. It is ready for receiving plants, and Cayley [*sic*] has the use of Gov't House at Parramatta to dry his specimens, &c'.[3]

Lieutenant-Colonel William Paterson (1755–1810), friend and correspondent of Sir Joseph Banks, Fellow of the Royal Society, was a competent amateur botanist who collected widely and established a fine garden on his estate at Petersham. He had been in the NSW Corps, acted as Administrator of the Colony, and had served as Lieutenant-Governor under Governor King. George Caley (1770–1829) was an energetic, if irascible, Banksian collector, with whom King had clashed on the voyage out in the *Speedy*. Although Caley chose to work at Parramatta, he was not impressed by plans for a botanic garden there. In December 1800 he dutifully reported to Banks: 'Governor King and Col. Paterson were some time back anxious to establish a Botanic Garden, but I hear nothing said of it now ... the ground is lying waste'.[4]

By the time the French naturalist François Péron visited Parramatta in 1802, the situation had changed:

The whole eastern front of Rose Hill ... is a very gentle declivity, on which appears the fine garden belonging to the government, in which many interesting experiments are made, with a view to naturalize foreign vegetables: here are also collected the most remarkable of the indigenous plants, intended to enrich the famous royal gardens of Kew. It is from this spot that England ... has acquired most of her treasures in the vegetable kingdom; and which have enabled English botanists to publish many important volumes. An enlightened botanical professor, who combines modesty and indefatigable exertion, had just arrived at the time of our visit to superintend the garden at Parramatta, and the learned colonel Paterson, to whom New South Wales is indebted for this establishment, has never ceased to take a lively interest in its success.[5]

Although Péron misunderstood the nature of the visit of the 'botanical professor', Robert Brown, who arrived in Sydney in May 1802 as botanist on the *Investigator* under Lieutenant Matthew Flinders, it is significant that he should have seen in the Parramatta garden the same activities as were carried out in the early government gardens in Sydney. As Péron's visit occurred when land at Farm Cove was still in private hands, it seems likely that he saw the Parramatta garden when it was the chief entrepôt and acclimatization centre.

Governor King referred to a 'Government garden in Parramatta' in 1806,[6] and twelve years later, Allan Cunningham reported to Banks an apparently new or revived venture: 'Governor Macquarie is forming a Botanic Garden at Parramatta under the direction of Charles Fraser, late Private, 46th Regt., who is styled Colonial Botanist'.[7]

In its last days, this Parramatta cultivation was tautologically known as the 'Horticultural Garden', apparently because Governor Brisbane had permitted the Agricultural and Horticultural Society of NSW to use the former government garden as a nursery and experimental plot. Faced with lively competition from private commercial nurseries, as well as from the Sydney Botanic Garden itself, the Society gladly offered the land to the Church Corporation in 1833, and Governor Bourke arranged for compensation of £300 to enable the Society to meet its liabilities. The King's School was established on the site, and the old Parramatta Garden disappeared forever.[8]

Despite, or because of, his valiant efforts to render fertile the rather sterile area at Farm Cove, Charles Fraser, as Colonial Botanist, wanted to establish a botanic garden on better soil, and far removed from any government vegetable patch. Accordingly, he prevailed upon Governor Macquarie to mark out an entirely new garden. The Governor agreed, no doubt seeing this as an additional safeguard of his privacy, and on 4 September 1821, he 'went by Water this forenoon accompanied by Mr Meehan Dy Surveyor Genl, Mr Chas. Fraser, Colonial Botanist, Cap.t Piper and Lieut. Macquarie—to *Double Bay* and there marked the future *Botanic* Garden; directing about Twenty acres of Ground to be reserved & located for that purpose'.[9] Macquarie later advised Earl Bathurst that he had marked out 'a large and suitable allotment of Ground (about 15 acres) on the South side of Port Jackson Harbour, two Miles from Sydney ... for a Colonial "*Botanical Garden*"...'.[10] By the end of 1821, this site on the western side of Double Bay was being cleared and fenced, and the Governor's friend, Captain John Piper, who had land in the vicinity, was gently warned not to 'interfere with the views or intended purposes of Government, more especially with the Government Botanic Garden intended to be established at "Double Bay", on all which reservations for Government, the Dep^y Surveyor General has received already my Instructions'.[11] However, Macquarie left the colony in February 1822, and the Double Bay proposal lapsed.[12] Brisbane made some compensation for this by ceding to the Garden part of the 'Governor's Demesne'.

As we have seen, during the next eighty years or so, successive Colonial Botanists, Superintendents and Directors were associated with several governmental and other institutions which had gardens and trees. The provision of plants and the deployment of labour to meet such increasing demands often taxed the resources of the Garden and its Campbelltown Nursery. One additional responsibility surpassed the others in both urgency and magnitude.

On Monday, 18 July 1887, Sir Henry Parkes, Premier and Colonial Secretary, drew Charles Moore's attention to the Act concerning the celebration of the colony's centenary. One provision was the resumption of 640 acres (one square mile) of land 'to be named Centennial Park'. Parkes directed:

You will without loss of time take formal possession of this land for the public purposes legally provided for and commence the work of laying out and planting the Park in accordance with the plan by Mr Surveyor Deering subject to alterations which may be agreed upon hereafter ...

The Lachlan Swamp near Randwick as it appeared in 1887 when chosen as the site for Centennial Park.
(National Library of Australia, Canberra)

The Park was to be ready for opening on Anniversary Day, 26 January 1888, and progress reports were to be furnished in the meantime. To meet this emergency, Moore was empowered to draft men, preferably married, from government relief works.[13]

Moore, who had recently returned from the Adelaide Jubilee Exhibition, certainly acted without loss of time. Accompanied by James Jones, Overseer of Domains, Moore visited the site on the day Parkes wrote his letter. Jones carefully recorded this excursion to 'the Lachlan Swamp a large water reserve near Randwick where it is proposed to lay out and create a new surburban park to be called the Centennial Park ...'. It was ironical that Charles Moore, who forty years before had battled to secure a reliable and adequate water supply for the Botanic Gardens, should now have the task of converting a former source of Sydney's water into a lake-studded recreational park.

The following day, Tuesday, 19 July 1887, Moore and Jones returned 'and took formal possession of the Lachlan Swamp on the part of the Government from the Corporation of Sydney...'. Jones proudly added: 'Personally drove the first stake & turned the first sod. The Centennial Park this day commenced & inaugurated.'

By September, Jones noted: 'over 400 men are employed. Clearing (shifting sand hills) quarrying & draining,' and shortly before Christmas: 'Mr Moore gave me formal charge of the *Centl. Park* as General overseer. I have been acting as such from the beginning also promised me a permanent appointment there ...'.

Jones graphically recorded work during the final crescendo of activity early in January 1888:

The year begins with little of importance in the Domains or Gardens; but great hurry bustle and eagerness in the Centennial Park trying to be ready for the grand official opening & dedication of same on the 26th inst. being our first Australian Centenary.

We have now been working at the Park since 19th July last. Clearing away the original bush & scrub Ploughing & breaking up ground & sowing & planting grasses Blasting rocks, levelling sandhills making roads &c &c.

Over 400 men are now employed here under Mr Moore's general directions. They are divided into about 15 gangs; besides piece workers, each gang overlooked by a non working ganger. These are to some extent regulated by an overseer, but as many of them are themselves inexperienced in this sort of work, including the overseer Mr Moore makes me largely if not wholly responsible for the laying out and economic working of the whole concern ...

Jones was accordingly involved in developing plans, placing guide-pegs, laying out roads and watercourses, acting as paymaster and as general inspector.[14]

At noon on Tuesday, 24 January 1888, the Centenary Celebrations were inaugurated when Lady Carrington unveiled the statue of Queen Victoria which (although in a different position) still stands in Queen's Square near St James's Church in King Street. It was felt that this statue compensated for that lost in the Garden Palace fire, and 'fully 50,000 persons' witnessed the ceremony. Appropriately, the statue of the Queen faced that of her late Consort, Albert the Good, erected nearby at the northern entrance to Hyde Park in 1868. On 21 August 1922, the Prince's statue was 'placed in position on the bank opposite the Chief Secretary's Buildings' while work proceeded on the underground railway.[15] It has stood on this 'temporary site' near the conservatorium ever since.

At noon on Thursday, 26 January 1888, Charles Moore's great moment arrived. Centennial Park—'the People's Park' in the words of the Premier—was opened and dedicated before nearly 50 000 people. The Governor, Lord Carrington, who was credited with the suggestion to turn the reserve into a park, duly officiated. It was a bright day, clear and warm, with a welcome cool sea breeze. The speakers were commended for their extreme brevity; the crowds cheered, although 'the vast majority ... could ... hear nothing of the words which were being said'; the four guns saluted; 2300 troops marched past—and some trees were planted.

The tree planting was reported as 'the most exciting part of the proceedings', for the viceregal party and other guests, chiefly governors of the other colonies, were mobbed by the 'good-humoured and well-behaved' spectators, who nearly sabotaged the operation through their enthusiasm. Endeavouring to keep clear the area around each tree as it was planted, the police were seen on at least two occasions, 'very busy, clearing a ring around trees which had already been planted, and the subsequent discovery embarrassed them not a little'.[16]

The trees, 'some thirteen in all from the nursery of the Botanical Gardens, were laid out in nice order, and fixed in holes ready for the ceremony'. The band played 'lively music' while Lady Carrington, on being handed a spade by Mr Moore planted a Cook's pine, (now *A. columnaris*), followed by the Countess of Carnarvon, who 'with equal grace and skill' planted a silky oak, *Grevillea robusta*.

Finally all was accomplished, and it was suggested that this triangular plantation, approached from the gates near St Matthias's Church, would become known as 'the Centennial triangle'. Cards were fixed to the trees indicating their botanical names and the names of those who planted them. 'Special care is to be taken of them', wrote a *Herald* reporter, 'may they flourish!' Another reporter referred to the new reserve as 'that noble and new domain—the Centennial Park' which was 'a gift ... bestowed, in perpetuity, upon the people, the value and beauty of which will be more and more acknowledged as years roll by; a gift which, unlike many others, can never, by any amount of familiarity, grow stale and unprofitable'.[17]

James Jones, who does not appear to have featured in the news reports, recorded his own impressions. He felt that amid the 'great illuminations and fireworks, triumphal arches, exhibition flags of all nations, Naval & Military reviews ... public dinners, balls & banquets,' the planting of the first trees in Centennial Park was the 'most important of all'. The triangular piece of high ground had been 'rocky & bare' until blasted, levelled and topped with good soil, trenched and manured 'for the reception of the first memorial trees'. The area had been enclosed within a fence of wire, battens and tea-tree branches to protect the young trees 'from cold or scorching winds & from curious or meddlesome people', and 'a large military tent was erected for shelter & refreshments but only the aristocracy and their friends and a few workmen were admitted'.

Jones carefully plotted the position of each tree and noted with satisfaction that a plan of the park was displayed, 'painted & embellished by the Colonial Architect from my working plan which I lent him for the purpose'.[18] One wonders whether Jones, Moore and any members present of the army of workmen managed to suppress their mirth or to disguise their feelings of righteous indignation, when Sir Henry Parkes declared that this beneficent provision 'for the enjoyment and the health of the people' had been 'created as by the touch of a fairy's hand'![19] Jones also recorded that he had planted two additional trees, both *Ficus*, on behalf of Moore and himself. By 1 May, however, two of the original 'official' trees required replacement. Lord Carrington's tree was replaced by a Norfolk Island pine, which Jones 'planted ... with care ... as a memorial of my own labours as well as for Lord Carrington's memory'.

This Centennial Triangle is still evident, although Parkes's statue installed in 1897 has disappeared from its pedestal, and only eight trees—five of them *Ficus*—grow where fifteen were planted. The plot is adorned by a bed of *Canna* at one apex, an unlabelled monument, and the two fine Russian cannon, trophies of the Crimean War, which flanked Governor Bourke's statue when it stood close to the Garden Palace gates.

Plagued by rheumatism and sciatica, Moore found the pressure to have the Park ready in time very demanding. By February 1888, there may have been a reaction. James Jones certainly thought so:

Mr Moore seems to be almost overwhelmed with the extent & responsibility at his age (68) of this large work. His patience and temper are often overtaxed and his knowledge of engineering or surveying is limited. A great advocate of low wages, and governs by threatening and terrorising ...

Mr Jones had his own views on the expected performance of a workman: 'Excavating—In loose ground a man can throw up about 10 cubic yards per day, but

This photograph by Charles Kerry shows Centennial Park had begun to takes shape by 1892 as a recreational park studded with lakes.
(Royal Botanic Gardens Library)

in hard or gravelly soils 5 yds. will be a fair days work'. He based his calculations on the belief that: '19 cub. feet sand, 18 of clay, 24 of earth, $15\frac{1}{2}$ chalk & 20 gravel will each make a ton.' Nor did he shrink from such pre-metric mathematical nightmares as presented by the task of checking the weight of blue-metal screenings from the Kiama quarries intended for Domain pathways:

1 square heap measuring as under
Length $20'10'' \times 23'1'' \times 5'8\frac{1}{2}''$ mean height and containing
114 Tons 8 Cwt @ £57-4-0.

$$24 \text{ cubic ft allowed to ton } \& \quad \frac{23'1'' \times 20'10'' \times 5'8\frac{1}{2}'''}{24} \quad = \quad \begin{array}{cc} \text{T} & \text{Cwt} \\ 114 & 8 \end{array}$$

—but Mr Jones was not satisfied, so he had several cart-loads weighed and took averages, finally calculating that the heap contained 105 tons $3\frac{1}{2}$ hundredweight. Little wonder Moore put him in charge of operations at Centennial Park. It was sad that they should have fallen out at the beginning of June 1888 when it was discovered that six months before, two men had been paid £2-5-0 instead of 25 shillings. One man repaid the excess pound, the other had left. Moore insisted that Jones and the time-keeper make up the deficit. Jones refused, pointing out that during the year he had paid the men £23 706-7-1 and had never been a penny out. Moore 'threatened me with suspension' and 'probable dismissal', recorded Jones, who was obliged to give in on the matter before leaving Sydney on 4 June on his 'annual visit of inspection, pruning and renovation of Gardens at various ry. [railway] stations'.[20]

The celebration of the second centenary of European settlement was in sight when the next major 'outside' development of the Botanic Gardens was undertaken. It was an appropriate development in every way—geographically, geologically, scenically, climatically, historically and botanically. In November 1804, George Caley penetrated an area of the Blue Mountains near the Grose River calling it the 'Devil's Wilderness'.[21] A little to the east he struggled up a basalt-capped eminence he named Fern Tree Hill, and in November 1823, Allan Cunningham climbed the same hill noting that 'the summit ... is named by the aborigines Tomah'. The rich brown basaltic soil supported stands of 'lofty species of *Eucalyptus*', including the fine tree now known as Brown Barrel or Mount Tomah Messmate, *E. fastigata*. Giant tree-ferns ('Tomah' to the Aborigines) *Cyathea australis* and *Dicksonia antarctica* abounded. The slopes were clothed with rainforest containing Coachwood, *Ceratopetalum apetalum* and sassafras, *Doryphora sassafras*. Many trees and shrubs were bound together by climbers including wild sarsaparilla, *Smilax australis* and Native Grape, *Cissus hypoglauca*, while the ferny ground-cover in the adjacent eucalyptus forest was splashed with yellow *Senecio linearifolius*. In both variety and luxuriance of plant growth, Mount Tomah was a botanical paradise which had fascinated George Caley, Archibald Bell and Surveyor Robert Hoddle, Allan and Richard Cunningham, Charles Moore, Louisa Atkinson, Thomas Cadell, William Woolls, J.H. Maiden and so many others.[22]

In March 1830, Governor Darling authorized a grant of two square miles (1280 acres) at Mount Tomah to Susannah Bowen (1776–1840), mother of George Meares' Countess Bowen (1803–89), who was credited with defining the boundaries of the County of Cook, comprising much of the Blue Mountains region. On 31 July 1883,

The relatively unspoiled bush environment of the Outer Domain was still evident when this photograph,
attributed to Alexander Brodie, was taken about 1870.
(Historic Photograph Collection, Macleay Museum, University of Sydney)

Bowen transferred this original grant and other Mount Tomah land to his son, George Bartley Bowen, who in August 1895 sold out for £7000 to Philip Charley of North Richmond.

Charley subdivided and sold several blocks of the Mount Tomah land, including a summit block of some 70 acres. This was purchased in November 1934 for £450 in the name of Effie Jane Brunet, wife of Alfred Louis Brunet. The son of a farmer, Brunet was born on 8 July 1892 at Liglet, in the Department of Vienne, in Western France. Having studied horticulture in France and England, he migrated to Australia and became a British subject on 16 May 1921. On 26 November 1927 he married Effie Jane Silvester, and by the time the Mount Tomah property was acquired, he was a well-established nurseryman at West Pennant Hills.

At Mount Tomah, Alfred Brunet raised cold climate plants, including tulips, daffodils and rhododendrons for the cut-flower market, using laurel hedges to protect seedlings and maturing flowers. After developing and utilizing the Mount Tomah nursery and garden for some thirty years, Brunet, by then a friend of R.H. Anderson, considered the value such a property could be to the Botanic Gardens in providing a habitat for many plants which could not be successfully cultivated in Sydney. Mrs Brunet, who held title to the land, wholeheartedly endorsed the idea, and inspection of the property was invited. Several were made during the early 1960s by the Director, Knowles Mair, Superintendent Warwick R. Watson, Senior Landscape Designer Maurice A. Watson and other senior staff, and the potential of the property as an extension of the Sydney Gardens was quickly and enthusiastically appreciated.

Alfred Brunet died at West Pennant Hills on 26 November 1968, while fires raged on the Blue Mountains and discussions continued in Sydney. Further inspections confirmed the wisdom and generosity of Mr Brunet's intention, and the matter of acquisition was taken up with his widow. Early in March 1969 Mr Mair's formal recommendation went forward, to be ultimately endorsed by the Minister for Agriculture. On 7 August 1970 Mrs Brunet signed the transfer papers to surrender more than 68 acres of Susannah Bowen's original grant to Her Majesty Queen Elizabeth II for the sum of one dollar, and the land was duly reserved for Botanic Gardens purposes on 14 July 1972, just four years before Mrs Brunet's death. Since 18 December 1984 the land has been registered in the name of the Royal Botanic Gardens and Domain Trust.[23] Work proceeds on extensive landscaping and on the planting of cold-climate species, while retaining original stands of such trees as *Eucalyptus fastigata* and pockets of rainforest. It is proposed to feature wide ranges of conifers and rhododendrons and thematic plantings of southern hemisphere species.

The most recent extra-mural project was announced by the Premier, the Honourable Neville Wran, at the launching of the Gardens Springtime Programme on 22 September 1984. After several sites had been considered by senior Gardens staff and representatives of the Department of Environment & Planning, recommendations were made to the government. Consequently, about 470 hectares (1161 acres) encompassing a watercourse and some eucalypt woodland were reserved 4 kilometres west of Campbelltown to be 'developed and managed by the Royal Botanic Gardens of Sydney' as the Mount Annan Native Botanic Garden and Arboretum. This second satellite garden will cater not only for the cultivation of Australian plants and botanical research, but also for the recreational needs of the rapidly growing residential districts of south-western Sydney. Thus for the second time in a dozen years, the aspirations of J.H. Maiden in 1912 were further realized—land has been set apart for botanical purposes, 'away from the smoke of a large city', to some extent contrasting 'as regards its soil, with the natural barrenness of the Sydney garden', yet 'readily accessible'; a reserve which 'would not take the place of the Sydney institution', but which would 'fill a real gap in the State's requirements'. As early as 1897, Maiden had regretted that 'New South Wales did not possess a single arboretum of the first class'. Commendably, nearly ninety years later, a government has taken a new opportunity when the scientific need for an additional reserve was no less pressing than the commercial demand for additional space.[24]

Chapter 9

RETROSPECT AND PROSPECT

*I should simply like to make the point that a Botanic Garden of world standard, which this
I hope is, although it is in need of and undergoing improvement in most spheres, is not
a glorified park ... It is a scientific and educational institution set in surroundings of beauty
and catering for peaceful recreation.*

Dr L.A.S. JOHNSON, Director, 1982[1]

Governor Phillip's bid to secure for all time a great government reserve adjacent to the settlement he founded has long been admired. Yet even in the time of his immediate successors there were clear deviations from his purpose, and inroads were made for nearly two centuries until halted, it is hoped, by the 1980 Act.

In Macquarie's time and later, regrets were expressed that even the greatly reduced reserve restricted the development of a rapidly growing settlement largely dependent upon maritime enterprise. In 1819, long before domain land on east Sydney Cove was sold to raise funds for a new viceregal residence, young William Charles Wentworth lamented that 'Government House, and the adjoining domain' denied 'greater facilities for the erection of warehouses and the various important purposes of commerce'.[2] Nearly forty years later, the visiting political economist, William Stanley Jevons (1835–82) revived the issue:

... in the original laying out of Sydney a great mistake was made; a large extent of land surrounding Farm Cove extending thence to the high ridge of Hyde Park & including both the promontories of Fort Macquarie and Lady [*sic*] Macquaries Chair were reserved for parks or other public purposes. The whole of this would be extremely valuable as affording both wharves for marine trade & a good central position for the other trades; at present the main part of Sydney is much confined on the east side by this reserved land and shipping is driven to ... inferior wharfs [*sic*] ...

Jevons maintained that any reserve should have been located in the Rocks area, 'that steep & to some extent useless & objectionable part ... between Sydney Cove & Darling Harbour ...'.[3]

As already observed, the commercial utilization of Farm Cove was advocated yet again in 1912, when Government House itself was the scene of the Labor Government's 'eviction' of the Governor-General, Lord Denman, and formal takeover of the residence and grounds.[4]

Such alarming comments, proposals and actions, with intermittent suggestions that some roadways in the Gardens and Domains should become regular public thoroughfares, frequently raised the question of the purpose and need of the reserves. Certainly they had been intensely and diversely used.

The changing role was reflected in administrative arrangements. During the foundation years, the Gardens and Domains were under the direct control of the Governor, through a superintendent of agriculture or a gardener-cum-convict overseer or a Colonial Botanist. The Governor in turn reported to the Colonial Office. Then for thirty-five years, 1821–56, the Colonial Botanist, Superintendent or Director was responsible to the Colonial Secretary, with or without the intercession of a Committee of Management. With responsible government, the Departments, first of Lands and Public Works, then of Lands, assumed responsibility through the Director. In 1880, jurisdiction returned to the Colonial Secretary until 1908 when the Department of Agriculture began its seventy year association. Since 1978 control has been vested in the Premier's Department through the Royal Botanic Gardens and Domain Trust established by the Act of 1980.

While subjected to the pressures and demands of political, bureaucratic and public caprice and exigency, those charged with the management of the Gardens and Domains have never lacked guidance from declared observation, comment, criticism, suggestion, or indeed, appreciation. Some press references to the efforts of Fraser, the Cunninghams and their successors have been noted. Many other notices described the standing of the reserves in individual or popular esteem.

Surgeon Peter Cunningham, who made the first of five voyages to Sydney in 1819 referred to 'the delightful promenade round the government domain' where 'our Sunday pedestrians and fashionables' enjoyed 'the cool evening sea-breeze among the delightful scenery'. However, he lamented that by 1826: 'the domain, beautiful as it still undoubtedly is, has lost much of its attraction since being deprived of the kangaroos and emus seen, in Governor Macquarie's time, hopping and frisking playfully about ...'.[5]

Surveyor William Romaine Govett who first saw Sydney late in 1827, was similarly impressed:

After entering the Domain, the walk opens into the carriage road, which is shaded on either side with thick and luxuriant indigenous shrubs for some way, until it passes the north wall of the Botanical Garden, where a bay and harbour opens into view, offering a prospect of bright waters, and fanciful masses of rock fringed with a variety of beautiful evergreens.

He found the scenery 'singularly pleasant and refreshing, and, perhaps, peculiar only to that part of the world'. An additional delight was the great variety of insects 'of the most beautiful and brilliant order' which greeted the visitor with 'an incessant loud and shrill buzzing'.[6]

During his last two years Charles Fraser constructed some greatly appreciated walkways, one of which was 1600 yards long. Shortly before Fraser's death at the end of 1831, the newly established *Sydney Herald* was concerned about Governor Darling's provisions for increased vehicular traffic:

The Domain—This favourite spot, the only pleasant walk in Sydney, is now rendered comparatively useless to the pedestrian, by its being made a drive for carriages. The inner domain is still exclusive—from the public. What are the inhabitants to do? we think it hard, that without ceremony, they should be curtailed in the *small* enjoyments the town affords. The public at large suffers, but more particularly our juvenile population; we trust that this unfair mode of vitiating public property, will be quickly abolished.[7]

The 'Governor's Demesne' had become public property.

The visiting Quaker naturalist-missionary, James Backhouse, was more interested in practical matters than in recreational walks. Having joined Alexander McLeay, Justice Francis Forbes, Sir John Jamison, William Macarthur 'and some other gentlemen' to inspect James Busby's grapevines in February 1836, Backhouse, not given to lavishing praise, tersely recorded: 'The Sydney Botanic Garden is a fine institution; it is furnished with a good collection of native and foreign plants. Some of its Curators have ranked highly as men of science'.[8]

Louisa Meredith, who arrived in Sydney in September 1839, left a much more spirited and incisive record of the Domain:

... close to the town is the beautiful Domain, a most picturesque rocky promontory, thickly wooded and laid out in fine smooth drives and walks, all commanding most exquisite views ... It was our favourite spot; even after driving elsewhere out of town (for alas! the splendour of George Street had no charms for me) we generally made one circuit round the Domain, and as generally found ourselves the only visitors. It was unfashionable, in fact, not the proper thing at all, either to walk or drive in the Domain. It was a notorious fact, that maid-servants and their sweethearts resorted thither on Sundays, and of course that shocking circumstance ruined its character as a place for their mistresses to visit; the public streets being so much more select.

Strangely, Mrs Meredith experienced the same loneliness in the Gardens:

The Government Gardens are tastefully laid out round the sloping head of a small bay between the Domain and Government House, and contain ... a strange and beautiful assemblage of dwellers in all lands, from the tall bamboo of India to the lowly English violet ... much as I should dislike to dwell in Sydney, I left its beautiful gardens with great regret. Yet, will it be believed, that even these are very little frequented by the inhabitants? They are evidently, from some cause unknown to me (but doubtless nearly allied to the cause of the Domain's desertion), not considered correctly fashionable by the fancied 'exclusives' ... though constantly frequented by all new-comers; at all events, the former prefer the hot, glaring, dusty pavement of a town street for their promenade, to these delicious gardens.[9]

A little later, Mrs Meredith acknowledged some remarks in a Sydney newspaper which suggested 'a more general resort to the Domain' than she had witnessed.

Christopher P. Hodgson, an amateur naturalist briefly attached to Leichhardt's first expedition, arrived in Sydney in 1839. He felt that in the Government Domain: 'Nature appeared to have concentrated her riches and rarities ... and the spirit of science and enterprize to have exerted their powers to render it a second paradise'.[10]

John Hood, who visited NSW in 1840–41, also appreciated the paradisiac qualities of the Gardens, while drawing some delightful allusions at the expense of the regulations:

I have just returned from the most beautiful spot I ever saw—the Botanical Gardens of Sydney. It was literally a walk through Paradise; the only difference betwixt it and Eden being, that here EVERY tree was forbidden, and death and destruction awarded, by man-traps and other means, to those that touched their fruit. These Botanical Gardens in position are the finest in the world ... But the splendour of the plants, the trees, the flowers! Every production of the East is here; every plant, every fruit, every beautiful flower is to be seen in these gardens in the highest possible perfection. The intensity of one's admiration is almost painful ...[11]

Like many sojourners, Governor FitzRoy's cousin, Lieutenant-Colonel Godfrey Charles Mundy confessed to 'having no science', yet considered the Botanic Gardens 'a most creditable effort on the part of a young colony'. In the late 1840s, Mundy found

the Botanic Gardens, divided into two compartments; one laid out in formal squares, containing the floral produce of many widely distant lands, flourishing together here as they flourish nowhere else; the other more in the English pleasure-ground style, embracing a wide circuit of the picturesque Farm Cove.

He condemned the 'rabid attack ... by the opposition members of the Legislative Council' upon the estimates for 1849, when in an attempt to embarrass Moore, 'this pleasant place of public resort ran imminent risk of being permitted to go to waste ...'. This 'disgraceful fact' was clearly an 'instance of radical ebullition and legislative wantonness'. Mundy agreed with Louisa Meredith that it was lamentable and incomprehensible that the ladies of Sydney seemed to prefer the dusty streets when like few 'towns of co-ordinate consequence' it was 'so bountifully supplied with breathing spaces close at hand.'[12]

As Mundy intimated, newly-arrived immigrants of the 1840s could soothe or arouse pangs of nostalgia by visiting the Gardens, where

the most valued plants are the English primrose, the cowslip, violet, and daisy, which are shaded from the sun by screens, and treasured as carefully as the most tender exotics ... in England. These simple and homely memorials of our native land touch the heart with their eloquent silence, and the sternest soul is not insensible to their mute appeal ...

even to the point of being moved to tears.[13]

For some people, like Lieutenant John Henderson, who stayed briefly in Sydney in 1844, the Gardens appeared 'elegantly laid out', with 'many fine plants' and afforded 'an extremely pleasant and beautiful walk, to the few respectable inhabitants who indulge in that exercise'. In fact, there were 'few or no other objects worth mentioning'.[14]

Whether observers found the Gardens and Domains crowded or deserted, appreciation of their location and utilization was expressed in generally lavish terms throughout the remainder of the nineteenth century. The naturalist-painter, George French Angas, who arrived in Sydney in July 1845, considered 'the Government gardens and domain are the most usual resort for the inhabitants ... Nothing can be more delicious, during one of the hot days of summer, than to seek the deep shade in the sylvan recesses of these gardens, and occupy one of the numerous rustic seats ...'.[15] Later, when Secretary of the Australian Museum, Angas described the Domain as 'a beautiful park ... richly timbered, and laid out with carriage drives, which command a series of the most enchanting views ...'.[16]

As Joseph Phipps Townsend noted in the early 1840s, this idyllic picture did not necessarily last all the year: 'The domain ... is rather pretty when it is not burnt to a cinder, nor looking like a red-hot frying-pan,' whereas 'the Botanical Gardens are more than pretty, and there may be found plants from every part of the world, all healthy and flourishing ...'.[17]

Another writer of the 1840s, J.O. Balfour, a Bathurst settler, recorded:

The plan, site, and general arrangement ... are as creditable to those who first designed them, as their uniform good order, the cleanliness of the walks and beds, and progressing improvements are to the present managing committee ... The Sydney gardens present on Sunday afternoons a most animated scene; persons of all classes flock to them, and seem to forget the toils and cares of the past week, in admiration of the beauty and loveliness that surround them.[18]

Surgeon R.G. Jameson also noted the social value of the Gardens and Domains when recording his impressions in 1841:

On the afternoon of Sunday, the Domain and Botanical Gardens are crowded with people of every degree. There the artisan, carrying one or two children, enjoys, with his better-half, the recreation of a walk, whilst his senses are gratified by the colours and odours of many and various plants, and his veins cooled by the light winds ... There the spruce clerks and the younger employés of the government offices exhibit their best attire. Troops of boarding-school girls ... are led in snow-white procession through the mazy walks; and age, reposing sedately, passes in review the sports and pleasures of a younger generation. Such is the scene, without any admixture of coarseness or immorality ... It would almost convince us that a genial climate really has a tendency to soften and refine the character of social life.[19]

The Reverend John Dunmore Lang, not the easiest man to please, added his tribute:

To those who are addicted to botanical researches, or to those who, like myself, merely delight to contemplate the wonderful works of God, without being very inquisitive about the genus and species of each, the Botanic Garden and the romantic walks of the Government Domain ... cannot fail to afford a never-failing source of far higher gratification.[20]

William Howitt, a Quaker naturalist who arrived in Sydney in June 1854, considered the Domain 'a beautiful park, beautifully wooded' through which one walked

to the Botanic gardens ... possessing one of the most enchanting situations imaginable ... They are divided into two parts, a carriage-way running between them. The upper one is more expressly laid out as a botanic garden, and contains a very comprehensive collection of plants and trees, all ticketed. The lower portion is more in the fashion of a pleasure-ground. The ground is retained in grassed lawns, with seats under the trees, and is, naturally, the ground resort of the Sydney public.

Unlike most visitors, Howitt listed many of the trees growing in the gardens which he declared were 'conducted on the most liberal principles; the public has the freest admission, and seeds and plants are supplied gratuitously to all parts of the world on application. I brought away myself nearly a hundred kinds of seeds ...'. Elizabeth

A flower show in the Gardens, probably that staged by the Australasian Botanic and Horticultural Society in September 1853. From a drawing by W.G. Mason in The Australian Picture Pleasure Book, *Sydney, 1857.*
(National Library of Australia, Canberra)

Macquarie was credited with 'having procured the laying out of these beautiful and extensive reserves' and saving 'these lovely features of the neighbourhood from the havoc and desecration of speculation and private cupidity'.[21]

Some of the anonymously-produced accounts of the 1850s described botanical as well as auriferous treasures for the would-be gold-seeker:

The botanical garden ... forms part of the Domain, and is ... an object of great attraction. Here are all those plants and trees which cause one's imagination to take flight to distant parts; the air is loaded with a delicious perfume, and the only sounds ... are those of the sea ... and the constant song of the ... locusts ...

Then followed the interesting observation:

The Upper, or, as it is sometimes called, the Middle Garden, was planned as early as 1812 by Mrs Macquarie, and formed out of the wild bush, under the superintendence of Mr Alexander [*sic*] Fraser, botanist ... This garden is known by the magnificent ... Norfolk Island pine which adorns its middle walk.
   The lower, and more extensive and beautiful garden, separated from the upper by the walk from the fort (i.e. Fort Macquarie) along the Government bathing house, and occupying the circular sweep of ... Farm Cove, was not in existence when Lieut.-Gen. Sir Ralph Darling assumed the government ... in December 1825. The spot which now smiles in all the loveliness of a highly cultivated garden was then nothing but thick bush, and bare, barren rocks ...[22]

This, by the Governor's direction and 'the persevering skill' of Fraser and Richard Cunningham, was made 'to rejoice and blossom'. The Gardens were later 'much extended and beautified' under 'Mr James Anderson, botanist, who effected greater improvements ... than all the former botanists together'—a warm, if tardy tribute.

182

William Jessop's account of 1862 was more qualified:

There are ... the Botanical Gardens, divided into the scientific and the pleasure-grounds. The former, old in appearance, careworn, and not very large, promise something in the distant future; at present their treasury is small, and contains a variety of good things, but a multitude of bad. The latter are magnificent, delightful, and invaluable ...[23]

The novelist Anthony Trollope, who visited Sydney in 1871, believed that 'the people of Australia ... are laudably addicted to public gardens' among which he considered 'those at Melbourne ... are the most pretentious, and, in a scientific point of view, the most valuable ...' but the Sydney Gardens 'beat all the public gardens I ever saw,—because they possess one little nook of sea of their own' and had the advantage of adjacent 'open sea-spaces' for shark-free bathing.[24]

Writing in 1872, Sir George Smyth Baden-Powell considered that the Domain was 'a public park of which Sydney has ... cause to be proud' while the Botanic Gardens, 'open free all day' provided 'a *place de ressort* not to be excelled in any part of the world'. Here, in a place 'singularly aided by nature, and also by clever management' there were 'beautiful views ... over rocks, from under clumps of bamboos, out of long avenues, over stately trees ...' and 'an interesting collection of live birds'.[25]

In May 1876, in deference to the many visitors to 'the recent Metropolitan Exhibition', the *Town and Country Journal* published a view of the Lower Garden and a description of the Domain and Gardens which comprised the city's 'earthly paradise'. The area reclaimed from the sea had been

laid out in ponds, lawns, and flower beds in a most pleasing manner; the original rocky grounds have been taken advantage of, and most artistic effects produced; rustic bridges, spanning small streams, have been constructed; while the planting and general laying out have been judiciously executed, so as to combine shade, shelter, mazy walks, delightful vistas, and quiet retreats ...

There was no mention of Mr Moore's 'arrangement grounds' nor of his plant labels, but there was a brief acknowledgement of his advice and management which had brought the Gardens 'to their present most attractive and satisfactory condition'. A more generous tribute was paid to 'Mrs Macquarie ... a lady whose memory should be perpetuated' for 'she was not only instrumental in getting the gardens and Domain dedicated to the public, but personally planned most of the pleasant roads and walks in that part of the city, and had them made under her own directions ...'.

Early in April 1878, about the time the 'Thirty Years' Wall' was completed, the *Town and Country Journal* favoured its readers with another account, couched in language as flowery as the Gardens themselves, but not lacking in perception:

There are two objects for which every visitor to Sydney is expected to express his enthusiastic admiration—our harbour and our Botanical Gardens. If he does not do so, he is set down at once as an unmitigated barbarian. Even Anthony Trollope, who is not thought to have exhibited an undue desire to crack up Australia and Australians, was obliged to render a glowing tribute to the beauty of our gardens. 'For loveliness', he says, 'and that beauty which can be appreciated by the ignorant as well as the learned, the Sydney Gardens are unrivalled by any that I have seen.'

The reporter agreed with Jessop that once inside the 1873 gates on Fig Tree Avenue, 'everything has an old-fashioned look, but ... this part is to be transformed and

The Aviary in the Gardens, as drawn by S.T. Gill for J.W. Waugh's The Stranger's Guide
to Sydney, *1861.*
(National Library of Australia, Canberra)

modernised'. Moore's Regulations of 1848 still greeted visitors with a string of caveats relating to smoking, plucking flowers and other aspects of behaviour 'in this realm of Flora'.

In a major publication to mark the first centenary of European settlement, Francis Myers described Sydney's Gardens with enthusiasm:

By far the most beautiful and highly improved of all our public reserves is the Botanic Garden, which is devoted to the development of the floral beauties of the temperate and semi-tropical zones ... a better site for a permanent botanic garden could hardly have been selected had the country been scoured for a dozen miles around ... Nature furnished a happy opportunity, and the gardener's skill has done the rest ...
Scientific botany has not been neglected in the gardens. There is a small museum containing a good and well-arranged collection, while for the benefit of students plants and trees are described by their botanic titles ...

Before the reclamation programme, 'the tide rose to the point where Allan Cunningham's monument now stands, and the walk round to the Governor's bathing house was a bit of rough rocky foreshore, thick with seaweed ...'.[26]

At the end of the century, William Dymock published an *Illustrated Guide to Sydney* which described the Gardens as

... extremely beautiful. They are in shape somewhat of the form of a crescent, and are clothed with flowers and plants from almost every known part of the world ... The Upper Garden is laid out for the most part in straight walks and square divisions. Being the oldest and best protected, it contains the greatest number of species ... The Lower Garden is now in a very complete state ... Unlike the Upper Garden, with its straight walks and regular compartments, every part of this is laid out in a more or less irregular form, clumped with plantations, with the larger growing trees dotted over the extensive lawns, composed of the 'couch' ... or the 'buffalo grass' ... Considerable attention has been paid to the classification of plants, by grouping various tribes together, by this means affording the botanist facilities for study, and the public an opportunity of extending their botanical knowledge ...[27]

Among the early twentieth century tributes was that paid by James Webb in his *Guide to the City of Sydney* ... (1905). The 'most charmingly situated' Gardens comprised three sections:

The Lower ... laid out in a multiplied variety of forms, dotted here and there with miniature lakes, plantation clumps, and compact verdant lawns.
The Upper and Middle gardens ... have been composed for the most part into straight walks and mathematically precise division. They comprise the older portion of the cultivation area, and owing to the better protected position ... they have been planted with the less hardy species of exotic plants and shrubs. In the original designing and arrangement of the Botanic Gardens generally, the happiest taste has been invariably displayed, and ... the greatest care and attention have been always lavished upon their up-keep ...

The Gardens, with seven hot-houses, museum and herbarium, afforded 'every facility to the student of horticulture to prosecute his researches ... the Botanic Gardens have acquired a world-wide fame, and their name has become a synonym for all that is beautiful ... derived from the marriage of art with native splendour ...'. *Guides* produced by the NSW Bookstall Company during the next thirty years and more retained much of this description, including the tribute to the original planning.[28]

The 'scenic emphasis' was long maintained. In 1963, the *Australian Encyclopaedia*, published in Sydney, advised:

The 'senior' Botanic Garden of Australia, that of Sydney, is distinguished chiefly for its picturesque situation ... ; in this setting its trees, shrubs, and lawns, though below the standard of Melbourne, are shown to marked advantage, and as the site is central the garden is an air-well much favoured by city people ...

The further point that 'perhaps the most impressive of Australian botanic gardens is that of Melbourne' did not go unnoticed in the parliamentary debate in February 1980 when it was hoped that 'in future years the *Australian Encyclopaedia* will be able to alter its opinion and say that not Melbourne's gardens but Sydney's are best ...'
The 1983 edition did not make comparisons or express opinions, beyond noting that within the Sydney Gardens, 'The National Herbarium of New South Wales ... contains a large collection of botanical specimens and a fine botanical library and is the chief seat of research on the systematic botany of New South Wales ...'. Thus the fires of botanical parochialism were quietly quenched.

Visitors entering the Gardens from Fig Tree Avenue in 1884 had the immediate choice of admiring the conifers,
the statuary or the contents of a huge hot house.
(National Library of Australia, Canberra)

While constant admiration of the picturesque location, beautiful views and pleasant walks was fully justified, it also revealed the limited view many commentators held concerning the purpose and significance of a botanic garden and their expectations from such an institution. Thomas Henry Braim, the perceptive headmaster of Sydney College, 1842–45, drew attention to the wider issues. Not only did the Gardens and Domains provide an 'extensive and beautiful promenade' but also a complete 'rus in urbe', a secluded retirement ...

The throwing these gardens open to the public serves many useful purposes: not only does it afford a pleasant promenade, but it induces a love for botanical science; it inspires an attachment to Nature's works; it leads ... the youthful mind to seek a more intimate acquaintance with those beauties which here cannot but be admired.[29]

As we have seen, statements were made, especially during economic and other crises, about the role of the Gardens, but the popular appeal of the place lay principally in its display of beauty and offer of sanctuary. These attractions tended to obscure the strictly scientific activities being rather covertly yet faithfully pursued in virtually closed buildings standing in one corner. The days of 'open access' were yet to dawn.

J.H. Maiden had been Director for barely a year when he delivered the presidential address in May 1897 to mark the seventy-sixth anniversary of the Royal Society of NSW. He advocated, among other things, the establishment of a 'real garden of trees', an arboretum, preferably within forty or fifty miles of Sydney with a small branch establishment in one of the colder regions; the punishment by 'a draconian law' of 'purveyors of false seed'; the conservation of timber trees, and greater care in thinning forests and in ring-barking; the promotion of botanical teaching; the establishment of 'an Institution for Botanical Research' and the conduct of a botanical survey of the State 'as correlative to geological survey'. To this end Maiden saw a grand exercise in co-operation with the University, Technical College, Technological Museum, government departments, and other institutions, and wanted 'the educational opportunities which the Gardens afford exercised to their fullest extent'. This, he believed, could be done 'without interfering with the ordinary work of the Gardens'.[30]

Fifteen years later, Maiden declared his views on the nature and function of botanic gardens in general. Initially avoiding mention of pleasant walks and aesthetically satisfying lawns and gardens, Maiden declared:

I think it may be fairly said that few people have more than a superficial idea of the activities of a modern botanic garden. Many people, and educated ones too, look upon my sub-department as having simply the horticultural and disciplinary care of a garden (and some parks), and think that the prefix 'botanic' is simply given as an explanation of the occurrence of labels on plants ...[31]

The fact had to be faced that many visitors 'are not botanical students primarily. So in Sydney we thickly coat the botanical pill with the sugar of a "garden of pleasure" ...'. Consequently

the average citizen, walking through his botanic garden, enjoys himself. He chats with his friend, he reads his book, he contemplates the changing scene of landscape, observes the manners and customs of his fellow men, or he rests,—simply approximates as far as he can to the idea of 'doing nothing', and thereby relieves jaded mind and body. Or, he takes his walking exercise along the paths or across the lawns, imbibing health by activity under pleasant extraneous conditions.

That is how the majority of visitors to a botanic garden occupy themselves. They take their botany mildly. Some of them imbibe a little in spite of themselves ...

A small minority ... are seriously interested in plants from the botanic or scientific aspect. The number is increasing ... and will increase, as facilities are given for teaching the subject, but let us be quite honest, and admit that the vast majority of the taxpayers are not botanists at all.[32]

While Maiden considered it beyond 'the range of practicability that the citizens at large will ever be freely admitted into the National Herbarium', he would be delighted to find today's more enlightened citizens identifying their own specimens in the Public Reference Collection, while others were enjoying guided tours or otherwise availing themselves of the specialized services now offered.

Today, as he nears the end of a long and distinguished career at the Gardens, the present Director, Dr Lawrie Johnson, recalls the great satisfaction he derived from seeing the formation of the Royal Botanic Gardens and Domain Trust; the completion and opening of the new herbarium building and its adjuncts; the establishment of a Division of Community Relations to complement the opening of the visitor centre; the provision of funds with which to develop the Mount Tomah garden; the proclamation of the second reserve at Mount Annan. Such developments reflect a keener and more practical interest in the Gardens than most governments have shown in the past.

Significant administrative changes have also been instituted. In mid-1985 Dr Barbara Briggs is Director's Deputy and Senior Assistant Director in charge of the Scientific Division; Mr Don Blaxell is Assistant Director in charge of the Living Collections Division; Mr Warwick Watson is Assistant Director in charge of Gardens Services and Mount Tomah; and Ms Susan Crick is Assistant Director in charge of the Community Relations Division. Mr Gordon Keen is currently in charge of the Secretariat and is Secretary to the Trust. The many and diverse facets of the organization's activities, ranging from the landscaping and planting of the grounds to the identification of plants, the maintenance of the Library and the organization of school visits, all come under the direct jurisdiction of one of the Assistant Directors, each responsible to the Director.

In the future Dr Johnson would like to see an exhibition building constructed to provide adequate and appropriate accommodation for horticultural exhibitions, thereby reviving one of Maiden's ambitions of seventy years ago. He hopes that the Visitor Centre will be enlarged to cater for even more ambitious displays, and that the work of the Community Relations Division, including the education section, will be further developed. In the grounds there should be an enrichment (in terms of species) in all gardens, and further informational plantings designed on thematic lines to demonstrate Aboriginal and European uses for certain species as well as aspects of plant ecology, adaptation and evolution.

At the research level, Dr Johnson is keen to see the long tradition of the study of eucalypts maintained; further studies made of relationships within the Australian flora, and of its relationship with the world flora. Systematic work must be continued, together with a vigorous publications programme, and the Gardens should retain involvement in ecological studies and current environmental issues through fieldwork, research, review and advice. In short, Sydney's Royal Botanic Gardens should serve as the main centre of enquiry and information relating to the State's plant species, their identification, classification, distribution and special significance.

It is believed that such aims will be best achieved under the leadership of one who is primarily a scientist, who although perhaps not specially trained in administration or in horticulture, must have an understanding of, and a sympathy for both, always with regard for community needs and expectations.

Over 150 years ago, Dr John Henderson predicted that the botanist and the politician, working together, would 'in the future history of the country ... be recognized as its benefactors, in proportion to their respective exertions, when their measures have reached maturity, perhaps long after the termination of their limited periods of existence'.[33] Charles Fraser, the Cunningham brothers, Charles Moore and Joseph Maiden would doubtless be bewildered by the political, technological and professional expertise required to attain these goals. Yet they would surely identify with the spirit and intention, whether the Gardens were coping with 100 000 visitors to a May Gibbs 'Snugglepot and Cuddlepie' Exhibition as happened in 1984, with a host of garden-lovers taking the Spring Walk as happens annually, with an enquiry about an algae-infested water supply as happens periodically, with requests to help resolve urgent botanical, horticultural or environmental problems as happens frequently, or with an individual citizen seeking to identify a solitary specimen as happens almost daily. They would, in James Backhouse's guarded words, still find 'a fine institution'.

Sources

Specific sources, both primary and secondary, are fully identified in footnotes to the text, where the importance of Annual and other reports will be readily appreciated. Most sources quoted are from published and unpublished material in the following repositories: the Library of the Royal Botanic Gardens, Sydney; the Mitchell Library within the State Library of New South Wales; the State Archives Office of New South Wales and the National Library of Australia, Canberra.

More general secondary sources include the following:

Anderson, Robert Henry, *An ABC of the Royal Botanic Gardens, Sydney*, Sydney, 1965.

Froggatt, Walter W., 'The Curators and Botanists of the Botanic Gardens, Sydney' in *JRAHS*, vol. XVIII, part 3, Sydney, 1932.

Gilbert, Lionel Arthur, 'Botanical Investigation of New South Wales, 1811–1880', unpublished Ph.D. thesis, University of New England, Armidale, 1971.

Gilbert, Lionel Arthur, 'Plants, Politics and Personalities in Colonial New South Wales' in D.J. and S.G.M. Carr (eds.), *People and Plants in Australia*, Academic Press, Sydney, 1981. (See also *JRAHS*, vol. LVI, part 1, 1970.)

Maiden, Joseph Henry, *A Guide to the Botanic Gardens, Sydney*, Sydney, 1903.

Maiden, Joseph Henry, 'History of the Sydney Botanic Gardens', part 1 (edited by R.H. Cambage) in *JRAHS*, vol. XIV, part 1, Sydney, 1928; part II (edited by W.W. Froggatt) in *JRAHS*, vol. XVII, parts 2 and 3, Sydney, 1931.

Mair, Herbert Knowles Charles, 'The Royal Botanic Gardens, Sydney: Birthplace of the Agriculture and Horticulture of a Continent', *Agric. Gaz. NSW*, April, 1970.

Tanner, Howard and Begg, Jane, *The Great Gardens of Australia*, Macmillan, Melbourne, 1976.

Wilson, Edwin (ed.), *Royal Botanic Gardens Sydney*, Royal Botanic Gardens and Domain Trust, Sydney, 1982.

Abbreviations

ANL	Australian National Library, Canberra.
AO	NSW Archives Office, Sydney.
ML	Mitchell Library, Sydney.
RBG	Royal Botanic Gardens Library, Sydney.
PRO	Public Record Office, London.
CSIL	NSW Colonial Secretary In-Letters.
MF,FM	microfilm
Gov. Desp.	NSW Governor's Despatches.
HRA	*Historical Records of Australia*, Series I.
HRNSW	*Historical Records of New South Wales.*
JRAHS	*Journal of the Australian Historical Society* (1901–18) and *Journal of the Royal Australian Historical Society* (from 1918).
ADB	*Australian Dictionary of Biography.*
SG	*Sydney Gazette.*
SMH	*Sydney Herald* (1831–42) and *Sydney Morning Herald* (from August 1842).
Aust.	*Australian.*
Mon.	*Monitor* and *Sydney Monitor* (from August 1828).
T & CJ	*Town and Country Journal.*

References

Introduction

1 *Annals of the Missouri Botanic Garden*, 2, February–April, 1915, p. 185.
2 Christopher Thacker, *The History of Gardens*, Sydney, 1979, p. 13 and *passim*.
3 Edward Hyams, *A History of Gardens and Gardening*, London, 1971, p. 9.
4 Thacker, op. cit., p. 8.
5 Hyams, op. cit., p. 121.
6 See for example, Thacker, op. cit., p. 127; Hyams, op. cit., pp. 128–9.
7 *cf.* R.T.M. Pescott, *The Royal Botanic Gardens Melbourne: A History from 1845 to 1970*, Melbourne, 1982, p. xi.
8 E. Hyams and W. MacQuitty, *Great Botanical Gardens of the World*, London, 1969, pp. 251, 253.

Chapter 1

1 Banks (Brabourne) Papers, vol. 3, ML A78^{-2}, p. 7.
2 Dr Wallis's reminiscences quoted in C. Annandale (ed.), *The Popular Encyclopedia*, London, 1885, vol. XII, p. 207.
3 L.A. Gilbert, 'Sir Joseph Banks' in *ADB*, I, 1966, and 'Cook's Companions: Young Mr Banks and Dr Solander' in *Armidale and District Historical Society: Journal and Proceedings*, 13, 1970.
4 J.C. Beaglehole, *The Endeavour Journal of Joseph Banks, 1768–1771*, Sydney, 1962, vol. I, p. 30.
5 op. cit., vol. II, p. 58.
6 J.H. Maiden, *Sir Joseph Banks: the Father of Australia*, Sydney, 1909, p. 84.
7 W.B. Turrill, *The Royal Botanic Gardens, Kew*, Jenkins, London, 1959, pp. 23–4.
8 Quoted from *House of Commons Journal*, xxxvii, p. 311 in Beaglehole, op. cit., vol. II, p. 113.
9 Beaglehole, op. cit., vol. II, pp. 112–13.
10 P.G. Fidlon and R.J. Ryan (eds), *The Journal and Letters of Lt. Ralph Clark, 1787–1792*, Sydney, 1981, p. 1.
11 G. Mackaness, *Admiral Arthur Phillip: Founder of New South Wales*, Sydney, 1937, p. 82.
12 *The Voyage of Governor Phillip to Botany Bay ...*, London, 1789, p. 41.
13 op. cit., p. 47.
14 op. cit., p. 48.
15 P.G. Fidlon and R.J. Ryan (eds), *The Journal of Arthur Bowes Smyth: Surgeon, Lady Penrhyn, 1787–1789*, Sydney, 1979, p. 64.
16 P.G. Fidlon and R.J. Ryan (eds), *The Journal and Letters of Lt. Ralph Clark, 1787–1792*, Sydney, 1981, p. 93.
17 *The Voyage of Governor Phillip to Botany Bay*, p. 58.

Chapter 2

1 *The Voyage of Governor Phillip to Botany Bay*, London, 1789, p. 221.
2 P.G. Fidlon and R.J. Ryan (eds), *The Journal of Arthur Bowes Smyth: Surgeon, Lady Penrhyn 1787–1787*, Sydney, 1979, p. 57.

3 George B. Worgan, *Journal of a First Fleet Surgeon*, Sydney, 1978, pp. 9–10.
4 Phillip to Sydney, 15 May 1788, *HRA*, 1, p. 19.
5 Daniel Southwell, Journal, *HRNSW*, 11, p. 665.
6 Ross to Nepean, 16 November 1788, *HRNSW*, 1, (2), p. 212.
7 P.G. Fidlon and R.J. Ryan (eds), *The Journal and Letters of Lt. Ralph Clark, 1787–1792*, Sydney, 1981, pp. 262–8.
8 Thomas Watling, *Letters from an Exile at Botany Bay to his Aunt in Dumfries; giving a Particular Account of the Settlement of New South Wales*, Penrith (1794), p. 7, letter of 12 May 1794.
9 Phillip's Instructions, 25 April 1787, *HRA*, 1, p. 15.
10 Phillip to Sydney, 15 May 1788, *HRA*, 1, p. 24.
11 David Collins, *An Account of the English Colony in New South Wales …*, London, 1798, p. 7.
12 P.G. Fidlon and R.J. Ryan (eds), *The Journal of Philip Gidley King: Lieutenant, R.N. 1787–1790*, Sydney, 1980, p. 39.
13 Collins, op. cit., p. 7.
14 Collins, op. cit., p. xxviii.
15 Worgan, op. cit., p. 13.
16 Collins, op. cit., p. 64.
17 *The Voyage of Governor Phillip to Botany Bay*, p. 127; Phillip to Sydney, 28 September 1788, *HRNSW*, 1, (2), p. 189.
18 Worgan, op. cit., pp. 12–13.
19 Collins, op. cit., p. 64.
20 op. cit., p. 51.
21 op. cit., p. 148.
22 op. cit., p. 64.
23 *SG*, 1 January 1804.
24 Charles Grimes, Plan of Sydney, May 1800 in *HRNSW*, V, after p. 837.
25 Macquarie to Goulburn, 17 May 1814, *HRA*, VIII, p. 252; Macquarie to Bathurst, 4 April 1817, *HRA*, IX, p. 373.
26 Collins, op. cit., p. 469; J.H. Maiden in *Kew Bulletin*, 1906, p. 207; *ADB*, 1, p. 438.
27 See map by James Meehan, October 1807, in *HRNSW*, VI, after p. 366.
28 *SG*, 11 and 18 January, 1807.
29 op. cit., 26 July and 2 August 1807.
30 Fitz to Under Secretary Chapman, 15 October 1807, *HRNSW*, VI, p. 305.
31 Harris to Anna Josepha King, 25 October 1807, *HRNSW*, VI, p. 347. 'Young Kent's' tombstone was discovered during excavations near old Government House in 1847. It bore the name of John Hunter Kent who died 18 May 1796, *SMH*, 16 June 1847.
32 *SG*, 15 July 1804.
33 op. cit., 16 and 23 March 1806.
34 King to Lord Hobart, 9 May 1803, *HRA*, IV, pp. 223–4.
35 F. Péron, *A Voyage of Discovery to the Southern Hemisphere …* London, 1809, p. 275.
36 J.H. Maiden in *Kew Bulletin*, 1906, p. 207; R.H. Anderson, *An ABC of the Royal Botanic Gardens*, Sydney, 1965, p. 21, but see also E. Wilson (ed.), *Royal Botanic Gardens, Sydney*, Sydney, 1982, p. 9.

Chapter 3

1 J.T. Bigge to Earl Bathurst in *Report … on the State of Agriculture and Trade in … New South Wales*, London, 1823, pp. 93–4.
2 Macquarie to Bathurst, 27 July 1822, *HRA*, X, pp. 671–2.
3 Government and General Order, *SG*, 6 October 1810; *HRNSW*, VII, p. 429.
4 Articles of Agreement, 21 August 1810, CSIL 1810, Bundle 4, pp. 90–105, AO4/1725.
5 M.H. Ellis, *Lachlan Macquarie: His Life, Adventures and Times*, Sydney, 1952, pp. 340–1; also *HRA*, X, p. 687.
6 *HRNSW*, VII, p. 634; see also Wentworth Papers ML A763, p. 211 and ML D1, p. 168; *SG*, 22 November 1817.

7 *HRA*, X, pp. 684–701.

8 Government and General Order, *SG*, 7 September 1811.

9 Government Public Notice, *SG*, 17 October 1812.

10 Colonial Letter Book (Copies of Letters Sent), 2 November 1813, AO 4/3491.

11 op. cit., Campbell to Palmer, 24 June 1814, AO 4/3493.

12 Macquarie to Bathurst, 7 October 1814, *HRA*, VIII, p. 341.

13 Gipps to the Marquess of Normanby, 23 December 1839, *HRA*, XX, pp. 441–6. In addition to Palmer's mills, there was one established by John Boston near the present Macquarie Street gates and Phillip's fountain, and another built by the competent millwright, Nathaniel Lucas, and later known as Kable's mill, in the vicinity of the present Sir Leslie Morshead fountain opposite the State Library. See Norman Selfe, 'Some Notes on the Sydney Windmills' in *Journal of the Royal Australian Historical Society*, Pt 6, 1902–03, pp. 96–107, and J.F. Campbell, 'Historical Notes on Government House Domain, Sydney' in *JRAHS*, XVII, Pt 2, 1931, pp. 111–26. Early in 1820, Thomas Williams received £12 'for removing the Windmill from Government Domain', *SG*, 29 July 1820.

14 J.T. Campbell to Geo. Crossley (agent for N. Divine), 8 April 1812, Colonial Letter Book, AO 4/3491; *HRNSW*, VII, p. 354.

15 Macquarie to Bathurst, 27 July 1822, *HRA*, X, p. 687. It is likely that the materials issued in November 1812 were for painting this 'Paling'; Macquarie: Letters (Addenda) ML A1932, p. 10a.

16 Nominal Return of Male Convicts Employed by Government at Sydney, 28 June 1813. Enclosure No. 7 with Despatch No. 1 of 1813, Macquarie to Bathurst, PRO, Colonial Office, Series 201. MF Reel 31. See also *HRA*, VII, p. 749. Flood was granted a free pardon in February 1813 (AO 4/1849 and 4/4427) and Alford returned to England a free man in 1817 (AO 4/1737).

17 Government and General Order, *SG*, 15 December 1810; *HRNSW*, VII, p. 468.

18 Blaxland to Macquarie, 4 December 1810, *HRNSW*, VII, p. 465.

19 G. Blaxland, *A Journal of a Tour of Discovery across the Blue Mountains*, Sydney, 1913, p. 13.

20 Government and General Order, *SG*, 10 June 1815; *HRA*, VIII, p. 569.

21 Blaxland to Banks, 10 November 1816 in W.R. Dawson (ed.), *The Banks Letters*, London, 1958, p. 105.

22 George Evans, Journal, ML MSS 589 and *HRA*, VIII, pp. 165–77.

23 Bathurst to Macquarie, 18 April 1816 and 30 January 1817, *HRA*, IX, pp. 114–15, 203.

24 Bathurst to Macquarie, 18 April 1816, *HRA*, IX, pp. 115–16.

25 Government Public Notice, *SG*, 6 July 1816.

26 Macquarie, Diary, ML A773, pp. 24–5.

27 Macquarie's Memoranda, 1814, ML A772, p. 41.

28 op. cit., p. 46.

29 J.M. Antill in *SMH*, 30 September 1897.

30 J.H. Maiden in *Kew Bulletin*, 1906, p. 209 (from *SMH*, 21 April 1906). See also J.H. Maiden in *SMH*, 16 and 23 June 1917.

31 War Office Records, London, WO 25/395; 12/5810.

32 Bathurst to Macquarie, 1 October 1816 (received 11 March 1817), *HRA*, IX, p. 188.

33 Banks's original draft has 'by Mr Aiton' crossed out. The Kew curator, William Townsend Aiton (1766–1849) reminded Banks on 29 May 1814 that before his illness George III had promoted botanical exploration in order to enrich the Kew Gardens. Aiton requested that this work be resumed. On 27 August 1814, Cunningham, no doubt on Aiton's advice, applied to Banks for employment as a botanical collector abroad. Dawson, *Banks Letters*, pp. 11–12; Banks Papers, ANL. G026 (MF).

34 Banks Papers, ANL. G026 (MF). See also NSW Col. Sec. Copies of Letters, AO 4/5782, pp. 383–6.

35 Banks Papers, ANL. G026 (MF).

36 ibid.

37 ibid.

38 Cunningham to Banks, 2 January 1817, loc.cit.

39 ibid.

40 Cunningham to Banks, 27 August 1814, loc.cit.

41 Draft instructions in Macquarie's hand, in CSIL (Surveyor General), AO 4/1814, pp. 7–9.

42 Macquarie's list of 'names and designations' of Oxley's party, 24 March 1817; John Oxley, *Journals of Two Expeditions into the Interior of New South Wales ...*, London, 1820, p. 362.

43 Cunningham's Journal, 28 April 1817, in Ida Lee, *Early Explorers in Australia*, London, 1925, p. 194.

44 Oxley to Macquarie, from Lewis Creek, Lachlan River, 28 April, 1817, Oxley Papers, ML. MSS. 589.

45 Cunningham's Journal, 30 May 1817, in Lee, op. cit., p.222.

46 Cunningham's Journal, 3 April 1817, in Lee, op. cit., p.174.

47 Cunningham's Journal, 28 April 1817, in Lee, op. cit., p.194.

48 Oxley to Macquarie, 30 August 1817; Macquarie to Bathurst, 5 September 1817; *HRA*, IX, pp. 477, 484.

49 Cunningham's Journal, 1 September 1817, in Lee, op. cit., p. 300.

50 Cunningham's Journal, 8 September 1817, in Lee, op. cit., p. 305.

51 Cunningham's Journal, 18 September 1817, in Lee, op. cit., p. 306.

52 Macquarie to Cunningham, 18 September 1817, Banks Papers, ANL. G026 (MF).

53 Cunningham to Banks, 1 December 1817, Banks Papers, ANL. G026 (MF).

54 Cunningham's Journal, 2 December 1817, in Lee, op. cit., p. 308.

55 Macquarie to Bathurst, 15 December 1817, *HRA*, IX, p. 729.

56 J.T. Campbell's enclosure with above letter, *HRA*, IX, p. 731. The bulbs would have been of the Darling, Murray or Macquarie Lily, *Crinum flaccidum* and of the Garland or Wilcannia Lily, Robert Brown's *Calostemma purpurea* which Cunningham synonymously named 'in honour of ... our worthy and much respected Governor ...', Cunningham's Journal, 30 April 1817, in Lee, op. cit. p. 197.

57 Cunningham to Banks, 20 December 1817, Banks Papers, ANL. G026 (MF).

58 Macquarie to Banks, 18 December 1817, Banks Papers, ANL. G026 (MF).

59 Banks to Macquarie, July 1818, Banks Papers, ANL. G026 (MF). The allusion is to the platypus, *Ornithorhynchus anatinus* Shaw.

60 Banks to Cunningham, 6–7 August 1818, Banks Papers, ANL. G026 (MF). The discrepancy must have been small, for Cunningham's journal records some 518 species and the number of Fraser's specimens sent to Bathurst was 'upwards of 500'.

61 ibid.

62 Macquarie to Bathurst, 30 May 1818, *HRA*, IX, p. 808.

63 Macquarie to Bathurst, 7 February 1821, *HRA*, X, p. 401.

64 Oxley, *Journals*, p. 209.

65 Macquarie's Instructions to Oxley, 9 May 1818, Oxley Papers, ML. MSS-5322, no. 48, p. 54.

66 CSIL, 1818–26, AO 4/1818, p. 1.

67 Oxley, *Journals*, p. 243.

68 Oxley, *Journals*, p. 387.

69 Macquarie to Bathurst, 23 and 25 March 1819, *HRA*, X, pp. 84, 136, 138.

70 Goulburn to Macquarie, 17 July 1820, *HRA*, X, p. 317; see also p. 136.

71 Goulburn to Macquarie, 24 March 1820, *HRA*, X, p. 297.

72 Macquarie to Bathurst, 18 July 1819, *HRA*, X, p. 177.

73 Macquarie to Bathurst, 22 July 1819, *HRA*, X, p. 195.

74 Macquarie to Bathurst, 7 February 1821, *HRA*, X, pp. 401–2.

75 Macquarie to Captain Skinner, 5 February 1821, CSIL, AO 4/1748, pp. 231–3.

76 Public Notice in *SG*, 3 July 1819.

77 Cunningham's Journal, 2 October 1818, in Lee op. cit., p. 403.

78 Cunningham's Accounts, December 1817–December 1818, Banks Papers, ANL G026 (MF).

79 *HRA*, IX, pp. 883–6.

80 Macquarie to Goulburn, 15 December 1817, *HRA*, IX, p. 735. See also J.T. Bigge: Report, Evidence, ML.B.T. Box 2, pp. 599–665.

81 J.T. Bigge, *Report of the Commissioner of Inquiry on the State of Agriculture an Trade in the Colony of New South Wales*, London, 1823, pp. 1, 3.

82 Bigge, op. cit., pp. 93–4.

83 Bigge, Appendix, ML.BT., Box 21, pp. 3662–81. 'List of Plants, Cultivated in Government Garden, Sydney, by Chas, Fraser, Colonial Botanist.', c. 1820. The lists include 'Plants cultivated from Seeds Received from Europe' (5 pp.), and 'from India' (5 pp.), a list of bulbs, a list of Van Diemen's Land Plants, and a long list of 'Australian Plants' (7 pp.) from such places as Port Macquarie, Mount Seaview, Peel Range, Blue Mountains, Liverpool Plains, Hastings River, Lake George, Cox's River, Bathurst, Macquarie River, Sutton Forest, Castlereagh River, Arbuthnot's Range (Warrumbungle Mountains), Fish River, Lachlan River, and undefined localities in the 'interior' and 'N.W. Interior'. The genera include *Eucalyptus* (17 spp.), *Angophora* (1 sp.) *Leptospermum* (6 spp.), *Cassia* (6 spp.), *Hakea* (4 spp.), *Grevillea* (1 sp.), *Banksia* (1 sp.), *Solanum* (2 spp.), *Hibiscus* (3 spp.), *Melaleuca* (3 spp.), *Casuarina* (7 spp.), 'Cupressus' (i.e. *Callitris*) (3 spp.), *Acacia* (8 spp., including Myall, *A. pendula*). Rosewood and cedar trees are also listed in this catalogue of some 126 species. Classifications of many species are incomplete.

84 Government and General Order, 6 January 1821, in Auditor General Appointments, AO 2/812, p. 115; *SG*, 6 January 1821; Macquarie to Bathurst (Enclosure) 30 November 1821, *HRA*, X, p. 581; Governor's Despatches, ML. A820, p. 112.

85 Returns of the Colony for 1822 ('Blue Book') p. 69.

86 Fraser's Petition of 4 September 1821, CSIL, Memorials, 1821, AO 4/1826, no. 48.

87 see chapter 8.

88 Lachlan Macquarie, *Journals of his Tours in New South Wales and Van Diemen's Land, 1810–1822* Public Library of NSW, Sydney, 1956, p. 223; Ellis, op. cit., p. 506.

89 *HRA*, XI, pp. x–xi.

Chapter 4

1 Colonial Office Miscellaneous Letters NSW, ML A2146, p. 85.

2 Macquarie, *Journals of Tours*, pp. 242–4; J. Jervis in *JRAHS*, XXVIII, 1942, pp. 299–300.

3 Fraser to Goulburn, 14 January 1822, CSIL, 1821–22, Bundle 16, AO 4/1752, p. 26.

4 *Proc. Roy. Soc. NSW*, LV, 1921, pp. lxxxv–lxxxvi.

5 Brisbane to Bathurst, 18 March 1825, *HRA*, XI, pp. 548–9.

6 op. cit., p. 548.

7 Stewart Murray to Charles Fraser, 30 June 1829, Letter Book, 1828–47, RBG A1.

8 Brisbane to Bathurst, 18 March 1825, *HRA*, XI, p. 587.

9 *SG*, 8 December 1825; Letter Book, 1828–47, RBG A1; McLeay to Fraser, 23 February 1829, Col. Sec. Letters to Estab. AO 4/3718; J.H. Maiden in *Public Service Journal*, 10 September and 10 December 1902.

10 *SG*, 10 July 1823; 3 October 1825; 26 May 1827. One quarry was 'near the beach in the Government Domain, contiguous to Woolloomoolloo'.

11 *SG*, 27 January 1825.

12 op. cit., 19 November 1827.

13 op. cit., 2 January 1827; 9 July 1828.

14 op. cit., 4 August 1828; 3 March 1829.

15 op. cit., 23, 25, 30 January 1828.

16 op. cit., 18 and 21 April 1828.

17 *SMH*, 28 November 1831.

18 Gov. Desp., ML. A1269, pp. 945, 957; ML. A1267. pt. 5, p. 532.

19 Gov. Desp., ML. A1208, pp. 953; *HRA*, XVI, pp. 179–81.

20 Bourke to Goderich, 2 November 1832; *HRA*, XVI, pp. 785–6.

21 Bathurst to Darling, 14 December 1825; *HRA*, XII, p. 88.

22 Darling to Under Secretary R.W. Hay, 11 January 1828, PRO Col. Office, Series 201, MF Reel 159. Only the Governor's covering letter has been found in PRO and ML records.

23 RBG, B1, pp. 446 *et seq.*; 450 *et seq.*

24 Goderich to Bourke, 22 February 1832, *HRA*, XVI, p. 526.

25 RBG, B1, pp. 387 *et seq.*

26 Quoted in full by J.H. Maiden in *Public Service Journal*, 10 January 1904, p. 10.

27 Gov. Desp., ML. A1209, pp. 699–700.

28 RBG, B1, p. 402.

29 *Public Service Journal*, loc.cit.

30 Gov. Desp., ML.A1209, pp. 699–700.

31 *SG*, 26 September 1827.

32 *Aust.*, 12 Oct. 1827.

33 *Mon.*, 1 Aug. 1829.

34 J. Atkinson, *An Account of the State of Agriculture & Grazing in New South Wales* ..., London, 1826, p. 59.

35 See for example, *SG*, 3 July 1819; 16 September 1824; 12 and 19 May 1825.

36 *SG*, 14 February 1829.

37 A shipment for Melville Island in 1825 included seeds of cabbage, carrot, parsnip, melons, cucumbers, pumpkins, squash, radish, celery, cress, parsley and mustard, and plants of pineapple, sugar, banana, coffee, lemon, orange, loquat, prickly pear, marjoram, thyme, spearmint, peppermint, sage, 'Cochineil Cactus', 'Hyssop and Cactus for pickling'. List in Fraser's writing in CSIL, Melville Island, 1823–28, AO 4/1802, pp. 101–2. The King George's Sound settlers especially asked for 'some good potatoes', AO 4/3718.

38 Letter Book, RBG, A1.

39 Elizabeth Macarthur, March 1827, in S. Macarthur Onslow, *Some Records of the Macarthurs of Camden*, Sydney, 1914, p. 458.

40 Elizabeth Darling, Australian Flowers, c. 1830, ML. B1026.

41 RBG, B1, p. 76.

42 Letter Book, RBG, A1.

43 ibid.

44 *SG*, 2 June 1825; 13 and 20 September; 2 December 1826; 5 January and 26 November 1827; 28 August 1830. *Aust.*, 16 September 1826; 17 February 1827.

45 CSIL, 28 January and 15 February 1828, AO 4/1966.

46 Darling to Murray, 1 February 1829, *HRA*, XIV, p. 631 and 9 October 1829, *HRA*, XV, pp. 200–2.

47 Murray to Darling, 10 April 1830, *HRA*, XV, pp. 409–10.

48 Letter Book, RBG, A1.

49 Col. Sec. Letters to Estab., 11 December 1828, AO 4/3718.

50 Letter Book, RBG, A1; Col. Sec. Letters to Estab., 25 November 1829, AO 4/3718.

51 Letter Book, RBG, A1; Col. Sec. Letters to Estab., 5 August 1829, AO 4/3718.

52 E. Macarthur to her son Edward, 26 May 1832, in S. Macarthur Onslow, op. cit., p. 467.

53 Letter quoted by Maiden in *Public Service Journal*, 9 January 1904.

54 J. Henderson, *Observations on the Colonies of New South Wales and Van Diemen's Land*, Calcutta, 1832, p. 85. This may well have been the Dr Henderson who came with a letter of commendation from Robert Graham of Edinburgh addressed to Charles Fraser, 13 March 1829, Letter Book, RBG, A1.

55 Col. Sec. Letters to Estab., 15 August 1831, AO 4/3719; Letter Book, RBG, A1.

56 *SG*, 10 September 1831. Anson's Point has long been more generally known as Mrs Macquarie's Point, the site of Mrs Macquarie's Chair.

57 Letter Book, RBG, A1.

58 St John's Burial Register, 519/1831 (where age is shown as forty-five); *SG*, 26 February 1820, 24 February 1821, 5 February 1824, 28 June 1826; Gov. Desp. ML A1211, p. 1117; A 1231, pp. 496, 499. Richard Cunningham, letter of 9 February 1833 in W.J. Hooker, *Companion to the Botanical Magazine*, London, 1836, vol. II, p. 213.

59 *Mon.*, 30 December 1831.

60 *Aust.*, 30 December 1831.
61 Richard Cunningham in Hooker, loc.cit.
62 Gov. Desp., ML A1211, pp. 1105–7.
63 *SMH*, 9 January 1832.
64 Bourke to Viscount Goderich, 4 January 1832, *HRA*, XVI, pp. 501–2.
65 McLeay to Under Secretary R.W. Hay, 5 January 1832, Col. Office Misc. Letters, NSW 1832, ML A2146, p. 252. Mrs Elizabeth Macarthur was one of those who hoped Cunningham 'may have the appointment'.
66 A. Cunningham to R.W. Hay, 7 May 1832, op. cit., pp. 78–9.
67 R. Brown to R.W. Hay, 10 May 1832, op. cit., p. 22.
68 R.W. Hay to A. Cunningham, 12 May 1832, op. cit., p. 93.
69 Hooker to McLeay, 4 July 1832, Papers of Linnean Society of London, ML FM 4/2699.
70 Letter Book, RBG, A1.
71 A. Cunningham to R.W. Hay, 17 May and 10 July 1832, ML A 1246, pp. 80, 82.
72 A. Cunningham to R.W. Hay, 23 July 1832, op. cit., p. 90.
73 A. Cunningham to R.W. Hay, 10 July 1832, op. cit., p. 832.
74 op. cit., p. 84. Cunningham enthused over such introductions as the camphor laurel, olive, lemon, lime, Indian teak, banana, date palm, cotton, passionfruit, plum and custard apple—but the introduction of desirable plants was only part of a botanic garden's true function.
75 op. cit., pp. 56–87. The Linnean or 'sexual' system of plant classification had lost much ground in favour of 'natural' systems since the publication of A.L. de Jussieu, *Genera Plantarum secundum Ordines naturalis disposita*, Paris, 1789, which advocated a natural system followed by A.P. de Candolle in *Flore française*, Paris, 1805, and by Robert Brown in *Prodromus Florae Novae Hollandiae*, London, 1810.
76 op. cit., pp. 89–90.
77 op. cit., p. 85—note before Section 5. This memorandum must be that said to be 'not available' in *HRA*, XVI, p. 701. See also Gov. Desp. ML A1269, p. 567. Bourke was asked to use the memorandum to prepare instructions for Richard.
78 A. Cunningham to S. Marsden, 13 August 1832, Marsden Papers, vol. I, ML A 1992, p. 525.
79 Hooker to McLeay, 4 July 1832, Papers of Linnean Society of London, ML FM4/2699.
80 George Bennett, *Wanderings in New South Wales*, London, 1834, vol. I, p. 338.
81 Letter Book, RBG, A1; Col. Sec. Letters to Estab., AO 4/3720.
82 *SMH*, 16 July 1832; Transcripts of Missing Despatches—Bourke to Goderich, 2 November 1832, ML A1267-4, p. 506.
83 Col. Sec. Letters to Estab., AO 4/3720, letter of 2 June 1832.
84 op. cit., letters of 16 June and 3 December 1832.
85 CSIL AO 4/2171.2, letter of 31 January 1833; Col. Sec. Letters to Estab., AO 4/3723, letter of 30 October 1833.
86 CSIL AO 4/2171.2.
87 AO 4/2222.4.
88 Col. Sec. Letters to Estab., AO 4/3723; CSIL AO 4/2171.2.
89 CSIL AO 4/2172.2 and Col. Sec. Letters to Estab., AO 4/3723.
90 CSIL AO 4/2171.2; Letter Book, RBG, A1.
91 CSIL AO 4/2171.2, letter of 8 August 1833.
92 AO 4/2222.4, letters of 28 April, 16 June, 8 July 1834; AO 4/3718; AO 4/3722.
93 Letter Book RBG, A1, McLean's draft Report for July–December 1833; Cunningham's Reports for January–June 1833, Gov. Desp. ML A1211, pp. 1099–102; and for 1834, Gov. Desp. ML A1212, pp. 1337–8 and A1213, pp. 27–8.
94 R. Cunningham to W.J. Hooker, 25 January 1835, quoted by Maiden in *Public Service Journal*, 10 February 1903, p. 12.
95 Col. Sec. Letters to Estab., AO 4/3722, letter of 20 January 1835, and AO 4/2268.2, letter of 15 January 1835.
96 op. cit., letter of 23 February 1835.

97 T.L. Mitchell, *Three Expeditions into the Interior of Eastern Australia*, London, 1839, vol. I, p. 178.

98 op. cit., p. 195.

99 Glenelg to Bourke, 4 July 1835. Gov. Desp. ML A 1272, pp. 265–6; *HRA*, XVIII, p. 5; AO 4/3722, letter of 9 November 1835.

100 *HRA*, XVIII, pp. 255, 789.

101 Col. Sec. Letters to Estab., AO 4/3722, letters of 2, 8, 15, 21 April 1836. See also R. Cunningham's letter of 23 September 1834 in AO 4/2222.4.

102 Col. Sec. Letters to Estab., AO 4/3722, letter of 14 January 1837.

103 Col. Sec. Letters to Estab., AO 4/3722, letters of 23 February and 22 March 1837; AO 4/2343.1, letter of 25 March 1837.

104 *NSW Govt. Gazette*, 1 March 1837, p. 213.

105 R.N. Dalkin, *Colonial Era Cemetery of Norfolk Island*, Sydney, 1974, p. 30. For an unhappy version of the incident, with an allusion to McLean's unpopularity among the convicts, see J.F. Mortlock, *Experiences of a Convict* (ed. G.W. Wilkes and A.G. Mitchell), 1966, p. 64. Note also the reference to landscape work at Capertee, 'all planned by Mr John McLean, at one time head gardener in the Botanical Gardens, Sydney, and afterwards holding a government appointment at Norfolk Island, where unfortunately he was drowned'. M. Herman (ed.), *Annabella Boswell's Journal*, Sydney, 1965, p. 12.

106 AO 4/2343.1, letter of 16 February 1837; AO 4/3724, letter of 26 May 1837.

107 AO 4/3722, letter of 27 February 1837. Dubbed 'the fever ship', the *Lady McNaughten* arrived on 26 February 1837 with nearly 100 people ill—56 had already died from typhus on the voyage from Cork. *SMH*, 27 February 1837.

108 AO 4/2343.1, letter of 25 March 1837; *ADB*, 2, pp. 609–10.

109 AO 4/2343.1, passim; AO 4/3724, letters of 8 May, 2 June and 5 July 1837.

110 *NSW Govt. Gazette*, 15 June 1836, p. 451.

111 Minute Book I, 1836–63, minute of 7 June 1836, Australian Museum Library.

112 Australian Council of National Trusts, *Historic Homesteads of Australia*, Melbourne, 1969, p. 72, quoting W.E. Riley, 1830.

113 The first meeting, 9 June 1836, was attended by William Macarthur, George Porter and Robert A. Wauch; the second, 16 June 1836, by Sir John Jamison and J.V. Thompson; no one came to the third meeting fixed for 23 June 1836. Syd. Bot. Gardens, Misc. Papers, AO 4/7577; Minute Book I, 1836–63, Australian Museum Library.

114 AO 4/7577; AO 4/2303.3, letters of 7 and 22 July and 15 September 1836.

115 AO 4/7577.

116 AO 4/2343.1, letter of 27 June 1837; AO 4/3724, letter of 15 July 1837.

117 AO 4/3724, letters of 10 and 15 July 1837.

118 AO 4/3724, letters of 3 June and 2 August 1837.

119 AO 4/3724, letters of 15 August, 26 September, 10 November 1837; AO 4/2222.4, letter of 8 July 1834; AO 4/2389.2, letters of 11 March and 18 July 1837.

120 E.D. Thomson to A. Cunningham, 10 November 1837, AO 4/3224.

121 ibid.

122 Cunningham to Thomson, 15 November 1837, AO 4/2343.1.

123 AO 4/2343.1; AO 4/3724, letters of 10 April, 24 November and 4 December 1837.

124 AO 4/2434.5, Anderson's Report, 21 January 1839.

125 AO 4/2389.2, letters of 4 January 1838. See also Anderson's Memorial and the committee's support, *HRA*, XIX, pp. 237–8.

126 AO 4/3724; AO 4/2389.2.

127 Letter of 15 January 1838, quoted by J.H. Maiden in *Kew Bulletin*, London, 1906, p. 215. See also *SMH*, 21 April and 23 May 1906.

128 *SMH*, 29 January 1838.

129 Gov. Desp., ML A1218. Report, pp. 597–616. See also AO 4/2389.2.

130 Gipps to Glenelg, 14 May 1838. Gov. Desp., ML. A1218, pp. 590–1.

131 Gov. Desp., ML A1218, pp. 628, 630–1. See also AO 4/2389.2.

132 op. cit., pp. 621, 624.

133 op. cit., pp. 635–8.

134 *ADB*, 1, p. 446.

135 Gov. Desp., ML A1218, pp. 592–3.

136 R. Heward, *Biographical Sketch of the late Allan Cunningham*, London, 1842, pp. 137–8. For a scholarly modern biography see W.G. McMinn, *Allan Cunningham: Botanist and Explorer*, 1970.

137 *SG*, 7 December 1839. In May 1843 a *Herald* correspondent complained that 'no steps' had been taken to erect the Cunningham monument although it was 'years … since it was first proposed … ', *SMH*, 2 May 1843.

138 AO 4/2434.5, letter of 22 August 1839.

139 AO 4/2480.1, letter of 3 December 1841.

140 *SMH*, 23 April 1842.

141 Leichhardt to Dr Wm. Nicholson (Bristol), 17 May 1842 in M. Aurousseau (ed.), *The Letters of F.W. Ludwig Leichhardt*, Cambridge, 1969, II, p. 470.

142 *NSW Govt. Gazette*, 2 December 1840.

143 Minute Book I, 1836–63, Australian Museum Library; *NSW Returns*, 1841.

144 Stanley to Gipps, 4 November 1842, *HRA*, XXII, p. 343.

145 Gipps to Stanley, 20 June 1843, *HRA*, XXII, p. 793.

146 Alexander McLeay was Colonial Secretary from his arrival in January 1826 until January 1837 when Edward Deas Thomson, Bourke's son-in-law, replaced him amid charges of nepotism.

147 McLeay to Thomson, 14 June 1843, *HRA*, XXII, p. 794.

148 Leichhardt to Dr Wm. Nicholson (Bristol), 17 May 1842, in Aurousseau (ed.), *Leichhardt*, II, pp. 469–70; also Leichhardt to G. Durando (Paris), 23 June 1842, op. cit., p. 493. Actually the salary was reduced from £200 to £140. The Gardens then cost about £800 year, Gipps to Stanley, 20 June 1843, *HRA*, XXII, p. 794.

149 Aurousseau, op. cit., II, p. 470.

150 *SMH*, 12 July 1844.

151 AO 4/2560.2, letter and note of 3 and 4 June 1842; also minute of 8 June 1842 in Minute Book I, 1836–63, Australian Museum Library.

152 AO 4/2560.2, letter of 25 June 1842; AO 4/3725, letter of 7 July 1842.

153 Leichhardt to Scott, 20 June 1844, in Aurousseau, op. cit., II, pp. 769.

154 Gipps to Stanley, 20 June 1843, *HRA*, XXII, p. 794; Col. Sec. Letters to Estab., AO 4/3725, letter of 27 January 1844.

155 AO 4/2638.2, letter of 30 January, 1844.

156 Macarthur to unnamed correspondent – possibly Dr Charles Nicholson, 15 July 1844. Macarthur Papers, vol. 37A, ML A2933, p. 57; AO 4/3725.

157 Macarthur to Hooker, 5 August 1844. Macarthur Papers, vol. 37B, ML A 2933, pp. 2–4.

158 Letter of 15 July 1844. Macarthur Papers, vol. 37A, ML A2933, p. 57.

159 Macarthur to Hooker, 5 August 1844. Macarthur Papers, vol. 37B, ML A2933, p. 4.

160 Hooker to King, 28 December 1844. King Papers, ML A3599, p. 278.

161 Leichhardt to G. Durando, 23 June 1842, in Aurousseau (ed.), op. cit., II, p. 492.

162 Macarthur to Hooker, 5 August 1844. Macarthur Papers, vol. 37B, ML A2933, p. 4.

163 AO 4/2628.2, McLeay to Gipps, letters of 20 July and 9 August 1844; AO 4/3725, letter of 1 August 1844.

164 AO 4/2677.2, letter of 4 May 1845; AO 4/3726, letter of 21 January 1846, *passim*.

165 AO 4/3726; AO 4/3727; AO 4/2755.3 *passim*.

166 AO 4/3726, letters of 20 May and 30 October 1845; Letter Book RBG, A1, letter of 28 October 1845.

167 Hooker to McLeay, 31 March 1845. Papers of Linnean Society of London, ML FM4/2699.

168 Hooker to McLeay, 1 September 1845, op. cit.

169 Gipps to Stanley, 20 January 1846, *HRA*, XXIV, pp. 722–3.

170 ibid. Gipps had a final confession to make concerning the Garden. He had to inform Stanley that it had escaped his recollection that the Legislative Council had in fact approved a salary of £200 'in the event of a scientific Botanist being appointed'. Gipps to

Stanley, 20 April 1846, *HRA*, XXV, p. 22.

171 Hooker to A. McLeay, 1 September 1845. Papers of Linnean Society of London, ML FM4/2699. John Smith was Curator of Kew Gardens, 1841–64. James Kidd was born 1 August 1801 in Fifeshire. On 14 April 1830 he was tried at Perth, Scotland, for forgery and sentenced to 14 years. He was then a gardener, married with four children. He arrived in Sydney on the *Burrell* on 19 December 1830, and was sent to the Botanic Gardens on disembarkation. Kidd was granted a ticket-of-leave in 1837 and a conditional pardon in August 1843. *Convict Indents*, AO 4/4016, p. 28; *Register of Conditional Pardons*, AO 4/4442, pp. 163–4. The Committee found no fault with Kidd, in fact, members recorded their satisfaction with his services: CSIL, letter of 29 September 1847. AO 4/3178; Minute Book I, 1836–63, minute of 14 September 1847 supporting Kidd's protest against salary reduction, Australian Museum Library.

172 Gladstone to FitzRoy, 16 June 1846, *HRA*, XXV, p. 98.

173 Minute of Executive Council, 2 January 1847, Gov. Desp. ML A1242, p. 460; *HRA*, XXV, p. 327.

174 FitzRoy to Grey, 3 April 1847, *HRA*, XXV, p. 454.

175 Wm. Elyard to Kidd, 21 August 1847, AO 4/3727.

176 Macarthur to Nash, 17 September 1847, Macarthur Papers, vol. 37B, ML A2933, p. 296. For R.W. Nash, see *ADB*, 2, p. 278, and Rica Erickson: *The Drummonds of Hawthornden*, Perth, 1969, pp. 59, 79–81.

177 Grey to FitzRoy, 10 July 1847, *HRA*, XXV, p. 657.

178 Grey to FitzRoy, 15 September 1847, op. cit., p. 752.

179 Moore's version of the note in Minutes of Evidence taken before the Select Committee on the Botanic Gardens, *V & P NSW Leg. Council*, 1855, vol. I, p. 1176.

180 John Stevens Henslow (1796–1861), botanist, geologist and Vicar of Hitcham, Suffolk; Professor of Botany at Cambridge.

181 Moore's version of the conversation, in Minutes of Evidence, *V & P NSW Leg. Council*, loc. cit. However, Hooker allegedly continued: '... had it been almost any other person, I should have felt it my duty to have expostulated with the Secretary of State'.

182 Macarthur to Hooker, 11 February 1848, Macarthur Papers, vol. 37A, ML A2933, p. 168.

183 Macarthur to Lindley, 11 February 1848, Macarthur Papers, vol. 37A, ML A2933, p. 163.

184 Bidwill to P.P. King, 26 December 1847, *Proc. Roy. Soc. NSW.*, 1908, p. 92; Wm. Elyard to Bidwill, 24 December 1847, AO 4/3727.

185 FitzRoy to Grey, 11 February 1848, *HRA*, XXVI, p. 229. The salary was shortly to be reduced.

186 Col. Sec. Letters to Estab., AO 4/3729, letter of 28 October 1853.

187 George Macleay to Robert Brown, 7 August 1848, Macleay Corresp. Brit. Mus. (Nat. Hist.) from Mrs Lydia Bushell, Macleay Museum, Sydney University.

Chapter 5

1 Minutes of Evidence, Select Committee on the Botanic Gardens, 10 August 1855, *V & P NSW Leg. Council*, 1855, I, p.1156.

2 Under Secretary B. Hawes to Moore, 21 July 1847, *HRA*, XXV, pp. 704–5.

3 Macarthur to Lindley, 11 February 1848, Macarthur Papers, vol. 37A, ML A2933, pp. 163–4.

4 E. Deas Thomson to Bidwill, 27 January 1848, Col. Sec. Letters to Estab., AO 4/3727.

5 AO 4/3728.

6 Col. Sec. Memorandum to Moore, 7 March 1848, *V & P NSW Leg. Council*, 1855, I, p. 1136. Manuscript draft AO 4/2792.2, and copy AO 4/3727.

7 E. Deas Thomson to Committee, 7 March 1848, AO 4/7577.

8 *Botanic Gardens: Report for 1848.*

9 Letter of 21 February 1848, AO 4/3727 and of 25 October 1848, AO 4/3728. Also letter of 14 December 1848: 'it is intended that the entire charge of the Domain should devolve upon you'.

10 Moore to Committee, 22 April 1848, AO 4/7577, and to Col. Sec. 14 August 1848, AO 4/2792.2.
11 AO 4/2792.2.
12 *NSW Govt. Gaz.*, 23 May 1848.
13 AO 4/7577.
14 Minute Book I, minute of 22 April 1848, Aust. Museum Library.
15 Kidd's estimates for 1847, forwarded 13 February 1846, AO 4/2726.2.
16 Col. Sec. Letters to Estab., 24 January 1848, AO 4/3727.
17 Wm. Elyard to Moore, 22 February 1848, AO 4/3727. These ten men were: Darby Carr, John Dunn, Charles Gritton, John Kitchen, Wm. D. Bartley, Wm. Courtney, James Dean, Robert Thompson, James Ryan and Richard Purcel. They were all employed at 3 shillings a day, six days a week. Men's Wages Book, RBG, C1.
18 Turner to Col. Sec. 25 October 1848, CSIL, AO 4/2792.2.
19 Moore protested that this amount included freight on 'five cases of Plants for the Botanic Garden' and so only the passage money of £85 was to be repaid at £13 month. Letters of 27 January and 31 March 1848, AO 4/3727. See FitzRoy to Earl Grey, 11 February 1848, for £300 having been voted for Moore's salary, *HRA*, XXVI, pp. 228–9.
20 Macarthur to Hooker, 1 February 1849, Macarthur Papers, vol. 37A, ML A2933, p. 178.
21 A superseded term for a large and diverse sub-division of Dicotyledonous plants including such Families as Chenopodiaceae, Amaranthaceae, Casuarinaceae, Proteaceae, Lauraceae and Euphorbiaceae.
22 Moore tried painted labels, then cast-iron frames with glass panels, but these, owing to the 'moist and genial season' of 1855–56, permitted the formation of mould and the labels became illegible.
23 *Botanic Gardens: Report for 1848.*
24 AO 4/2836.4, where there is also a manuscript copy of this first Report.
25 Carron to Committee, 22 June 1849, AO 4/7577; Men's Wages Book, RBG, C1; for William Carron see L.A. Gilbert in *JRAHS*, vol. 47, pt 5, October 1961, pp. 292–311, and in *ADB*, 3, pp. 360–1. Carron had unsuccessfully applied for the Overseer's position in June 1846. Minute Book I, Aust. Museum Library.
26 *Botanic Gardens: Report for 1849.*
27 Letter of 4 September 1849, AO 4/2836.4; letter of 11 September 1849, AO 4/3728.
28 Turner to Col. Sec., 1 February 1850, with *Botanic Gardens: Report for 1849.*
29 Letter of 19 February 1850, AO 4/3728.
30 *Botanic Gardens: Report for 1851.*
31 *Proc. Roy. Soc. NSW*, 1896, p. 19.
32 Letter of 4 July 1850, AO 4/3728. For the voyages in the *Medway* (1847–48) and in HMS *Havannah* (1850), see C. Moore, *Diary*, ML B786.
33 *Botanic Gardens: Report for 1850.*
34 Letter of 18 January 1850, AO 4/3728.
35 Letter of 26 August 1850, loc. cit.
36 Letters of 18 and 28 February 1852, loc. cit.
37 Turner to Col. Sec. 17 November 1851, *V & P NSW Leg. Council*, 1855, 1, p. 1137; Minute Book I, minute of 15 November 1851, Aust. Museum Library.
38 E.D. Thomson to Committee, 22 November 1851, *V & P NSW Leg. Council*, loc. cit.; AO 4/3728.
39 Letter of 20 May 1851, AO 4/3728.
40 Letter of 27 January 1852, loc. cit.
41 Letter of 10 February 1852, loc. cit.
42 Letter of 22 April 1851, loc. cit.; *Botanic Gardens: Report for 1850.*
43 *Botanic Gardens: Report for 1851.* These major classifications have long been superseded.
44 ibid.
45 *Botanic Gardens: Report for 1852*; letter of 16 March 1852, AO 4/3728.
46 E.D. Thomson to Moore, 27 August 1852, Col. Sec. Letters to Estab., AO 4/3729.
47 Thomson to Moore, 7 September and 18 November 1852, loc. cit.

48 *Botanic Gardens: Report for 1853.*
49 op. cit. and *Catalogue: Paris Exhibition, 1855*, pp. 114, 123–5.
50 Letter of 25 August 1854, AO 4/3729.
51 Moore to Col. Sec. 19 July 1854, Sand Drifts, Newcastle, RBG QVF 631. 45.
52 *Botanic Gardens: Report for 1853.*
53 Mitchell to Turner, 3 July 1852, *JRAHS*, 1931, pp. 155–6.
54 AO 4/7577.
55 *V & P NSW Leg. Council*, 1855, I, p. 1158.
56 R.H. Anderson, *An ABC of the Royal Botanic Gardens, Sydney*, Sydney, 1965, p. 29.
57 *JRAHS*, 1931, p. 156.
58 *V & P NSW Leg. Council*, 1854, II, p. 1291.
59 Elyard to Moore, 21 October 1854, Col. Sec. Letters to Estab., AO 4/3729.
60 Moore to Col. Sec. 24 October 1854, *V & P NSW Leg. Council*, 1854, II, p. 1293.
61 T.W. Shepherd, son of Thomas Shepherd (1799–1835), founder of the Darling Nursery (1827), and Michael Guilfoyle, who established a nursery at Double Bay in 1851.
62 *V & P NSW Leg. Council*, 1854, II, p. 1295.
63 op. cit., p. 1296.
64 *V & P NSW Leg. Council*, 1855, I, p. 1151.
65 op. cit., p. 1170.
66 op. cit., pp. 1158, 1168.
67 For evidence and findings of the Select Committee, see *V & P NSW Leg. Council*, 1855, I, pp. 1135–79.
68 Elyard to Moore, 16 January 1856, AO 4/3729.
69 *Catalogue of Plants in the Government Botanic Garden, Sydney, New South Wales*, Sydney, G.P., 1857 (8vo. 36pp.) – revised editions issued 1866 (47 pp.) and 1895 (130 pp.).
70 Letter of 27 February 1856, AO 4/3729.
71 Letters of 12 February, 20 April, 17 May, 25 August and 8 September 1855, loc. cit.
72 Letter of 28 February 1854, loc. cit.
73 *Botanic Gardens: Report for 1855.*
74 *Botanic Gardens: Report for 1856.*
75 Men's Wages Book, RBG, C1.
76 *Botanic Gardens: Report for 1857.*
77 *London International Exhibition, 1862 ... Catalogue of the Natural Products of New South Wales ...*, London, 1862, pp. 27–32.
78 J.H. Maiden in *SMH*, 2 May 1905.
79 *V & P NSW Leg. Ass.*, 1861, II, p. 1001.
80 op. cit., p. 1044.
81 G. Bennett, *Gatherings of a Naturalist in Australasia*, London, 1860, p. 334.
82 The first three of these terms are still used to describe the mode of attachment of stamens in flowering plants.
83 *Sydney Mail*, 8 September 1860, 25 October 1862.
84 Moore to Macarthur, 21 August 1863, Macarthur Papers, vol. 41, ML A2397, pp. 371–2.
85 *Sydney Mail*, 16 June 1866.
86 *Transactions of the Philosophical Society of N.S.W.*, 1862–65, pp. 204–5.
87 *SMH*, 18 and 21 February 1867. Kidd died on 15 February 1867, aged sixty-six and was buried in St Stephen's Churchyard, Newtown.
88 Resolution of 20 September 1867, AO 4/7577.
89 *V & P NSW Leg. Ass.*, 1868–69, III, p. 145 *et seq.*
90 *Sydney Mail*, 23 July and 15 October 1870.
91 *Botanic Gardens: Report*, 24 May 1871, reprinted in *Sydney Mail*, 15 July 1871.
92 *Records of the Aust. Museum*, Sydney, vol. XI, 1916–17, p. 361.
93 Journal, Botanic Gardens, 1871–85, AO 5/4802.
94 *Sydney Mail*, 16 September 1871.
95 *Botanic Gardens: Report*, 29 March 1879.
96 *NSW Govt. Gaz.*, 7 February 1878.

97 *Official Record of the Sydney International Exhibition, 1879*, Sydney, 1881, p. xx.

98 op. cit., pp. xxi, xxv.

99 Labour Record Book, 1877–82, RBG, C13.

100 *SMH*, 23 September 1882. The actual number recorded was 1 117 536.

101 ibid.

102 Labour Record Books, RBG, C13 and C23.

103 *V & P NSW Leg. Ass.*, 1883, I, pp. 545 *et seq.*

104 *SMH*, 28 June 1883; 20 February 1884.

105 *SMH*, 8 December 1885.

106 *Daily Telegraph*, 10 December 1885.

107 J. Jones: Diary, AO 2/8558.

108 Mueller to Moore, 2 March 1887, AO 4/7577, pp. 125–8. Even so, Moore published some additional locality notes. *Proc. Roy. Soc. NSW*, 1893, pp. 84–5.

109 *SMH*, 12 May 1896.

110 Botanic Gardens and Branches: Register of Employees, AO X1600.

111 *SMH*, 12 May 1896.

112 F.W. Burbidge in *Journal of the Kew Guild*, 1905. vol. 11, p. 265.

113 *SMH*, 12 May 1896.

114 F.W. Burbidge in the *Gardeners' Chronicle*, 13 May 1905.

115 *Proc. Roy. Soc. NSW*, 1896, p. 19.

116 *SMH*, 2 May 1905; *Proc. Roy. Soc. NSW*, 1908. p. 114. For some personal details relating to Charles Moore I am indebted to his great-niece, Mrs D.L. Cox, of Leura.

Chapter 6

1 *Proc. Roy. Soc. NSW*, 1912, p. 56.

2 *Proc. Roy. Soc. NSW*, 1880, p. 10.

3 Maiden to Professor A.J. Ewart, 20 May 1921, Corresp. Files, Agriculture: Botanic Gardens, AO 8/262.

4 Banks Papers, vol. 16, ML, and AO 4/7577 (formerly ML A2638) with note by John Metcalf.

5 *Report for 1896–1897*. Subsequent unacknowledged quotes are from *Reports* for appropriate years. Space precludes their separate acknowledgement.

6 *Proc. Roy. Soc. NSW*, 1912, p. 58, and *Report for 1912*, p. 5.

7 A.H.S. Lucas in *Proc. Linn. Soc. NSW*, 1930, p. 364; Maiden to A.T. Clerk, 15 June 1922, AO 8/262.

8 *SMH*, 1905, newscutting in ML Q991/0, vol. 2.

9 *JRAHS*, 1931, p. 155; Maiden, *Sir Joseph Banks*, p. 41.

10 Maiden to Deane, 27 March 1909, ANL, MS 610.

11 A.H.S. Lucas, loc. cit.

12 Maiden to Deane, 18 July 1910, ANL, MS 610.

13 Aviary, Register of Birds, June 1896–October 1940, AO 5/4800.

14 *SMH*, 25 March 1912.

15 AO 8/262.

16 *SMH*, 14 June 1916.

17 Reminiscences of Mrs M. Colditz (1983) RBG.

18 Maiden to Deane, 11 February 1919, ANL, MS 610.

19 Maiden to Deane, 13 December 1920, ANL, MS 610.

20 AO 8/262.

21 A.H.S. Lucas, loc. cit.

22 *Sun*, 2 October 1924.

23 Maiden to Deane, 12 April 1915, ANL, MS 610.

24 *Sun*, loc. cit.

25 *Daily Telegraph*, 17 November 1925.

26 *Sydney Mail*, 20 June 1891; *Chemist and Druggist of Australasia*, 1 October 1893; *Public Service*

Journal, 9 February 1901, 1 June 1915, 15 November 1924 and 15 January 1925; *Aust. Journal of Pharmacy*, 20 June 1924; *Sun*, 2 October 1924, 16 November 1925; *Daily Telegraph*, 3 October 1924, 17 and 21 November 1925; *SMH*, 17 November 1925; *Evening News*, 16 November 1925; *Australasian* (Melbourne) 28 November 1925; *Argus* (Melbourne) 21 November 1925; *New Nation*, September 1926; *Gardeners' Chronicle*, 13 December 1924; *Kew Bulletin*, 1926; *Proc. Linn. Soc. NSW*, vol. 51, 1926 and vol. 55, 1930; *Proc. Roy. Soc. NSW*, vol. 60, 1926; *Proc. Roy. Soc.* (London) B, vol. 100, 1926; *Aust. Naturalist*, July 1930; Maiden's *Annual Reports*, 1896–1923, RBG; Personal information from Mrs Helen Nicholas and Mr Harry Craig, grandchildren of J.H. Maiden, and donors of material to RBG library.

27 *Agric. Gaz. NSW*, 1902, p. 195.

Chapter 7

1 Address at opening of the new Herbarium, 6 November 1982.
2 *NSW Govt. Gaz.*, no. 88, 4 July 1924.
3 *ADB*, 8, pp. 212–13; *SMH*, 11 June 1924, 13 April 1942; *Jour. Aust. Inst. Agric. Science*, vol. 8, 1942, pp. 85–6.
4 Valder to Darnell-Smith, 19 August 1924, and associated documents, AO 8/262.
5 *Annual Report*, June 1925.
6 Rollo Gillespie, *Viceregal Quarters*, Sydney, 1975, pp. 70–2; *Annual Reports*, June 1924 and June 1925.
7 *Annual Report*, June 1926.
8 AO 8/263.
9 *Annual Report*, June 1929.
10 *Annual Report*, June 1931; AO 8/263; for an interesting history of the Campbelltown Nursery, 1881–1930, see Verlie Fowler in *Grist Mills*, vol. 1, nos. 3 and 4, April and June 1983, and her full report, RBG.
11 AO 8/264A.
12 ibid.
13 AO 8/263.
14 AO 8/262, 8/263.
15 Memorandum of 26 May 1932, AO 8/263; *Annual Reports*, June 1933 and June 1934.
16 AO 8/264A.
17 Memorandum of 7 May 1931, AO 8/263.
18 AO 8/264A.
19 *SMH*, 26 August 1931.
20 AO 8/264A.
21 AO 8/262.
22 ibid.
23 AO 8/264B.
24 C.W. Peck to E. Cheel, 24 August 1935, AO 8/264B.
25 AO 8/264 A and B.
26 Personal reference from R.D. Watt, Professor of Agriculture, 18 October 1920. Anderson Papers in possession of Mrs Beverley Passey. See also J.W. Vickery's obituary of Anderson in *Contrib. N.S.W. Nat. Herb.*, vol. 4, no. 5, 1972; *Daily Telegraph*, 8 August 1953.
27 AO 8/264B.
28 ibid.
29 Submission written by Anderson on behalf of Under Secretary, (September 1945?), Anderson Papers in possession of Mrs Beverley Passey.
30 News cutting of 31 December 1936 in AO 8/264B.
31 *Annual Report*, 30 June 1954.
32 Anderson to Noble, 6 August 1958, Anderson Papers in possession of Mrs Beverley Passey, Armidale.
33 Anderson to Noble, 27 October 1958, loc. cit.; *Annual Report*; 30 June 1959.
34 *Annual Report*, 30 June 1961.

35 *Daily Mirror*, 22 June 1960.

36 *NSW Industrial Arbitration Reports*, 1962, pp. 250–306; *SMH*, 17 October 1962.

37 Alma T. Lee, Obituary of J.W. Vickery in *Telopea*, vol. 2, no. 1, 1980.

38 See *Agar in Australia*, CSIRO, Bulletin 203, Melbourne 1946.

39 *Telopea*, vol. 2, no. 3, 1981.

40 Alma T. Lee in *Telopea*, vol. 2, no. 1, 1980, p. 4.

41 *Daily Telegraph*, 12 March 1964.

42 Zelie McLeod in *Daily Telegraph*, 8 August 1953.

43 *Contrib. NSW. Nat. Herb.*, vol. 4, no. 5, 1972.

44 *Sun*, 29 September 1968.

45 For an appreciation of this interesting man, see R.C. Carolin and L.A.S. Johnson in *Telopea*, vol. 1, no. 2, 1976.

46 See W.R. Watson and A.N. Rodd, 'The Glass Pyramid', *Agric. Gaz. NSW*, April 1978.

47 H.K.C. Mair, 'The Royal Botanic Gardens—Birthplace of the Agriculture and Horticulture of a Continent', *Agric. Gaz. NSW*, April 1970.

48 Norman Hall, *Botanists of the Eucalypts*, CSIRO, 1978; supplement, 1979, pp. 5–6.

49 A.S. George in *Flora of Australia*, Canberra, 1981, vol. 1, pp. 6–7.

50 Personal communication, 6 December 1984.

51 *Annual Report*, 30 June 1971.

52 Personal communication, 6 December 1984.

53 Copy in RBG library.

54 *Annual Reports*, 30 June 1972, 1973, 1974, 1975, 1976, 1977.

55 For the debates see *NSW Parliamentary Debates* (Hansard), *Third Series, Session 1979–80*, Sydney, 1981, pp. 4368–71; 4861–9; 4925–38.

56 29 Elizabeth 11, 19 (1980).

57 The other members were Mr Antonio Fazzini, Mr John Edward Ferris, Mrs Irene Eva Morphett and Professor Michael George Pitman.

Chapter 8

1 *Proc. Roy. Soc. NSW*, 1912, p. 59.

2 *HRA*, I, p. 402.

3 King to Banks, 3 May 1800, *HRNSW*, IV, p. 82.

4 Caley to Banks, 22 December 1800, Banks (Brabourne) Papers, ML.

5 Péron, *Voyage of Discovery*, p. 283.

6 King to Castlereagh, 27 July 1806, *HRNSW*, VI, p. 113.

7 Cunningham to Banks, 19 December 1818 in W.R. Dawson (ed.), *The Banks Letters*, London, 1958, p. 247. It is likely that more than one garden was established at Parramatta. See Merle Thompson in *Native Plants for New South Wales*, April 1985, pp. 21–4.

8 S.M. Johnstone, *The History of the King's School, Parramatta*, Sydney, 1931, pp. 54–5; also Col. Sec. letter of 23 April 1849, *JRAHS*, 1928, p. 13.

9 L. Macquarie, Diary 1818–22, ML A774, p. 235.

10 Macquarie to Bathurst, 27 July 1822, *HRA*, X, pp. 688, 845.

11 Macquarie to Piper, 7 September 1821, CSIL, 1821, AO 4/1749.

12 see Henry Selkirk Papers, RAHS, M 28.

13 Parkes to Moore, 18 July 1887, AO 4/7577.

14 James Jones, Diary, AO 2/8558.

15 Note by J.H. Maiden in Dept. of Agric. Herbarium Correspondence, AO 8/262.

16 *SMH*, 27 January 1888.

17 *Syd. Illust. News*, 22 February 1888.

18 Jones, Diary.

19 *SMH*, 27 January 1888.

20 ibid.

21 see A.E.J. Andrews (ed.), *The Devil's Wilderness: George Caley's Journey to Mt. Banks, 1804*, Hobart, 1984.

22 *JRAHS*, 1939, pp. 498–9; Ida Lee, *Early Explorers in Australia*, p. 513; Keith Ingram in *Hawkesbury Flora*, Soc. for Growing Aust. Plants, Spring 1983 and Spring 1984.

23 Mount Tomah Memorandum, 1973, RBG; *SMH*, 27 November 1968, 22 July 1976; Registrar General's Records; personal information from Mrs Lorna Backhouse, Mount Tomah.

24 J.H. Maiden in *Proc. Roy. Soc. NSW*, 1897, p. 51; op. cit., 1912, p. 59; Premier's Dept, Press Release, 22 September 1984.

Chapter 9

1 Address at the opening of the new Herbarium, 6 November 1982.

2 W.C. Wentworth, *Statistical, Historical, and Political Description of The Colony of New South Wales ...*, London, 1819, p. 6. He expressed the same view in the 3rd edition, 1824.

3 W.S. Jevons, 'Remarks upon the Social Map of Sydney, 1858', ML B864; *SMH*, 9 November 1929.

4 Led by the Premier J.S.T. McGowen, and the instigator, the Deputy Premier, W.A. Holman, the symbolic act of taking possession occurred on Saturday, 14 December 1912. The matter finally reached the Privy Council in 1915. The Government's right to dispose of the property as it wished was recognized. In November 1915, Sir Gerald Strickland moved into the residence as Governor. See *SMH*, 16 December 1912; R. Gillespie, *Viceregal Quarters*, pp. 17, 230–5.

5 P. Cunningham, *Two Years in New South Wales ...*, London, 1827, I, pp. 39–40; 67.

6 W.R. Govett, 'Sketches of New South Wales' in *The Saturday Magazine*, London, 1836.

7 *SMH*, 6 September 1831.

8 J. Backhouse, *A Narrative of a Visit to the Australian Colonies ...*, London, 1843, p. 351.

9 L. Meredith, *Notes and Sketches of New South Wales...*, London, 1844, pp. 39, 41–2.

10 C.P. Hodgson, *Reminiscences of Australia ...*, London, 1846, p. 11.

11 J. Hood, *Australia and the East ...*, London, 1843, pp. 103–4.

12 G.C. Mundy, *Our Antipodes ...*, London, 1852, I, pp. 70–3.

13 G.F. Angas, *Savage Life and Scenes in Australia and New Zealand ...*, London, 1847, II, 190–1.

14 J. Henderson, *Excursions and Adventures in New South Wales ...*, London, 1851, I, p. 70.

15 G.F. Angas, op. cit., p. 190.

16 [G.F. Angas], *Australia: A Popular Account ...*, London, 1855, p. 149.

17 J.P. Townsend, *Rambles and Observations in New South Wales ...*, London, 1849, p. 158.

18 J.O. Balfour, *A Sketch of New South Wales*, London, 1845, p. 58.

19 R.G. Jameson, *New Zealand, South Australia and New South Wales...*, London, 1842, pp. 112–13.

20 J.D. Lang, *An Historical and Statistical Account of New South Wales...*, London, 1852, II, pp. 174–5.

21 W. Howitt, *Two Years in Victoria...*, London, 1860, pp. 252–5.

22 *A Visit to Australia and its Gold Regions*, London, 1859, pp. 121–2.

23 W.R.H. Jessop, *Flindersland and Sturtland...*, London, 1862, I, p. 89.

24 A. Trollope, *Australia and New Zealand...*, Melbourne, 1873, pp. 138–40.

25 G.S. Baden-Powell, *New Homes for the Old Country...*, London, 1872, pp. 26–7.

26 Andrew Garran (ed.), *Picturesque Atlas of Australasia*, Sydney, 1886, I, pp. 91–3.

27 *Dymock's Illustrated Guide to Sydney*, Sydney, 1899, pp. 34–7.

28 J. Webb, *Guide to the City of Sydney ...* Sydney, 1905, pp. 144, 146. See also revised edition, c. 1933, pp. 68, 70, and c. 1939, pp. 59–61.

29 T.H. Braim, *A History of New South Wales ...*, London, 1846, II, pp. 295–6.

30 *Proc. Roy. Soc. NSW*, 1897, pp. 51–69.

31 *Annual Report for 1912*, p. 24.

32 *Proc. Roy. Soc. NSW*, 1912, pp. 54–5.

33 John Henderson, *Observations on the Colonies of New South Wales and Van Diemen's Land*, Calcutta, 1832, p. 86.

Index

ABC of the Botanic Gardens   152
Aborigines   ix, 3, 8, 42, 188
Adiantum formosum   146
Agathis robusta   101
Agland, John   21
Agricultural and Horticultural Society, NSW   169
Agricultural Gazette of NSW   121, 126, 142
Agricultural Hall (Domain)   116, 120, 129, 134
Agriculture, Department of   x, xi, 129, 131, 139, 140, 142–144, 148, 151, 152, 154, 158, 161, 164, 178
Aiton, William   5
Aiton, William Townsend   28, 36, 50, 193
Alford, Diane   163
Alford, Thomas   13, 16, 21, 193
Anderson, James   56, 57, 61, 62, 65–67, 126, 182, 198
Anderson, Robert Henry   x, xii, 139, 145–149, 151, 152, 155–157, 159, 176, 189, 204, 205
Andersons, Andrew   166
Angas, George French   180, 181
Annan, Mount   176, 187
Antill, Major Henry   26
Anzac Bed (Garden)   136
Anzac Buffet   134, 135
Apsey, John   15
Araucaria columnaris (A. cookii)   172
Araucaria cunninghamii   101
Araucaria heterophylla   16, 26, 27, 34, 46, 93, 104, 147, 161, 162, 173, 182
Arboretum   164, 176, 186, 187
Armstrong, Captain Richard R.   108
Armstrong, W.   68
Arrangement Grounds   51, 60, 80–82, 84–86, 94, 97, 104, 117, 130, 151, 158, 163, 184, 185, 188
Art Gallery   111
Arundo donax   11
Astrolabe   11
Atkinson, Caroline Louisa Waring   174
Atkinson, James   45, 196
Augusta, Dowager Princess of Wales   5, 6
Australasian Botanic and Horticultural Society   80, 81, 182
Australian   44, 49
Australian Encyclopaedia   185
automobiles   127
aviary   98, 99, 103, 107, 130, 131, 183, 203
Backhouse, James   179, 188
Backhouse, Lorna   xii, 206

Baden-Powell, Sir George Smyth   183
Baker, Richard Thomas   116
Balfour, J.O.   181
Banks, Sir Joseph   4–7, 11, 16, 17, 21, 27–30, 32, 33, 117, 126, 166, 168, 191, 193, 203
Banks Memorial   126
Baptist, John   26, 90
Barker, Thomas   91
Barnet, James J.   105, 106
Bartley, William D.   201
Barwick, Sir Garfield   161
Bathing House, Governor's   143, 182, 184
Bathing Machine   53
baths   94, 96, 118, 120, 121, 123, 183
Bathurst, Henry, Third Earl   20, 22, 23, 26–28, 30, 31, 33–39, 43, 169, 192, 193
Bathurst   41
Baudin, Nicholas   16
Beagle   65
Beard, Dr John Stanley   xi, xii, 151, 160, 161
Beattie, Sir Alexander   xi, 154, 165, 166
Bedford, David   xii, 155
Bell, Archibald   174
Bell, Henry   104
Bengal Merchant   46
Bennett, Dr George   59, 66, 82, 97, 105, 113, 197
Bent, Judge Jeffrey Hart   37
Bentham, George   49, 89, 99, 108
Betche, Ernst   110–112, 118, 120, 129, 131
Bidwill, John Carne   69, 72, 73, 75, 76, 82, 87, 92
Bigge, Commissioner John Thomas   17, 38, 41–43, 192, 195
Blake, William   37
Blakely, William Faris   123, 137, 143
Blaxell, Donald F.   163, 165, 188
Blaxland, Gregory   21, 193
Blaxland, Mount   21
Bligh, Captain William, Governor   15–17
Boorman, John Luke   125
Bosisto, Joseph   100
Boston, John   193
Botanic Gardens Creek   11–16, 26, 54, 83, 88, 104, 121, 147, 160
Botanic Gardens Trust, *see* Royal Botanic Gardens and Domain Trust
Botanical Congress, 1981   163, 165
Botanical Miscellany   48

Botany Bay   5, 7–9
Botany Lectures   84, 87, 92, 93, 97, 98, 121
Boulton, Thomas   18
Bourke, Sir Richard, Governor   43, 50–53, 58, 59, 61–63, 84, 105, 118, 124, 169, 173
Boussole   11
Bowen, George Bartley   175
Bowen, George Meares Countess   174, 175
Bowen, Susannah   174
Bowman, Dr James   61
Brachychiton acerifolius   37
Braim, Rev. Thomas Henry   186
Breakwell, Ernest   131, 137
Bridges, Frederick   116, 117
Briggs, Dr Barbara   xii, 154, 155, 163, 188
Brisbane, Sir Thomas, Governor   39–41, 43, 170
Britten, James   126
Brothers   56
Broughton, Ven. (later Rt.Rev.) William Grant   61
Brown, Robert   36, 50, 74, 87, 126, 165, 197
Brunet, Alfred Louis   175, 176
Brunet, Effie Jane (née Silvester)   175, 176
Buchan, Alexander   5
Buffalo   53
Buring, Ida and Leo   26, 149
Burke, John   21
Burnett, Alexander   68
Burrell   200
Busby, James   51, 53, 54, 58, 63, 66, 179
Cadell, Thomas   174
Cahill Expressway   104, 107, 151–53, 156, 165
Caley, George   120, 168, 174
Callitris glaucophylla   89
Callum, James   13
Calostemma purpurea   32, 194
Calvert, Louisa, *see* Atkinson
Cambage, Richard Hind   122, 126, 143
Camfield, Julius Henry   107, 118, 120, 128, 137, 138
Campbell, John Fauna   122
Campbell, John Thomas   20, 22, 39, 193
Campbell, Walter Scott   166
Campbelltown Nursery, *see* Nursery, State
Camphor Laurel, *see Cinnamomum camphora*
Cannabis sativa   154
Carpet Bedding   124, 136

Carr, Darby   201
Carrington, Charles Robert, Lord (Governor)   172, 173
Carrington, Cecilia Margaret, Lady   171
Carron, William   82, 100, 111, 201
Carter, Norman   137, 162
Carters' Barracks   53
Centennial Park   xi, 111, 112, 117, 118, 120–122, 124, 131, 134, 143, 144, 148–151, 156, 161, 164, 170–174
Ceratopetalum apetalum   174
Chaffey, Frank   140
Charles II, King   4
Charley, Philip   175
Cheel, Edwin   120, 129, 131, 142, 145–147
Chippendale, George   x, xii, 150
Choragic Monument   149
Cicadas   178, 182
Cinnamomum camphora   46, 105, 145
Cissus hypoglauca   174
Clark, Charles   134
Clark, Lieut. Ralph   10, 191, 192
Clarke, Rev. William Branwhite   66, 67, 69, 70, 84, 107
Clianthus formosus   82
Coachwood, *see Ceratopetalum apetalum*
Cohen, Mrs D.   150
Cohen, K.   154
Colditz, Mrs M.   204
Colley, Alexander   34
Collins, Captain David, Judge-Advocate   192
Committee Of Enquiry (1855)   90–93
Committee Of Enquiry (1861)   95, 96
Committee Of Enquiry (1869)   100
Committee Of Superintendence (Management)   58–60, 64, 66–69, 75, 77, 79, 80, 82, 84, 85, 91, 92, 97, 117
Commonwealth, Inauguration of   123, 124, 126
Connor, John   103
Conservatories   x, 40, 86, 88, 104, 117, 145, 159, 165, 185, 186
Conservatorium Of Music   20, 43, 137, 171
Constable, Ernest Francis   150, 158
Contributions from the NSW National Herbarium   147, 149, 150, 155, 156, 163
Cook, James   111
Cook, Lieut. James   5, 6, 10, 101, 158, 160

Cooney, Patrick   21
Courtney, William   201
Coveny, Robert G.   158
Cowper, Charles   91
Cox, Dorothea L.   xii, 203
Cox, William   22, 39
Craig, Harry   xii, 204
Crawford, Geoffrey R.   158
Crick, Susan   188
Cricket   95–97, 118, 134, 144
Crinum flaccidum   32, 35, 194
A Critical Revision of the Genus Eucalyptus   125, 139, 141, 143
Crocodile   46
Crocodile (Ornament)   146
Cross, Dr Douglas O.   xii, 148
Cullen, Sir William   135
Cunningham, Allan   ix, 27–34, 36, 37, 40, 41, 50–52, 56–58, 60, 62–66, 68, 69, 84, 91, 120, 126, 169, 174, 188, 193, 194, 197–199
Cunningham, Surgeon Peter   178
Cunningham, Richard   50–57, 61, 84, 174, 182, 188, 196, 197
Cunningham Monuments   ix, 65, 87, 126, 184, 199
Cunninghamia   163
Curnow, James   67
Currie, Thomas   36
Cutler, Sir Roden, Governor   107
Cyathea australis   174
Cyperus rotundus   94, 167
Daniel, Sylvanus B.   95
Darius, King   1
Darling, Eliza   44, 46, 196
Darling, Lieut.-Gen. (later Sir) Ralph, Governor   42, 47, 174, 179, 182
Darling Nursery   58, 82
Darnell-Smith, Dr George Percy   139–146, 151
Dasyurus   129
David, Professor Sir T.W. Edgeworth   113
Davis, Margaret   161
Dawes, Lieut. William   12
Dean, James   201
Deane, Henry   116, 130
De Candolle, Augustin P.   86, 87, 197
De Jussieu, Antoine L.   111, 197
De La Martiniere, Bossieu   11
Delaney, Nicholas   21, 25
Dendrobium moorei   100
Dendrobium speciosum   34
Denison, Alfred   98
Denison, Sir William, Governor   90, 92, 95, 99
Denman, Thomas, Lord (Gov.-Gen.)   178
Dent, Samuel   21
Depression, Economic   68, 142–145
Devonshire Street Cemetery   66, 126
Dick, Alexander   96
Dicksonia antarctica   174
Divine, Nicholas   13, 16, 20
Dixson, Hugh   117
Dodd, Henry Edward   11, 13
Domain   ix, 2, 16, 18–21, 23–26, 37–39, 42, 44, 47–49, 52, 55, 57, 71, 78, 82, 84, 87, 95–97, 101, 103–106, 111, 117, 118, 121, 123,

127–129, 131–138, 143, 145, 147–153, 156, 159–161, 171, 174, 175, 177–184, 186, 195
Donaldson, Stuart Alexander   81, 91, 92
Doryanthes excelsa   34
Doryphora sassafras   174
Double Bay   39, 169, 170
Dracophyllum fitzgeraldii   100
Drinking Fountains   110, 118, 133, 150
Driver, Richard   96
Dromedary   17, 36
Duff, John   111
Dunlop, J.C.   108
Dunlop, Robert   104
Dunn, John   201
Dymock, William   185
Earl Of Pembroke   5
Eden, Garden Of   1
Edwards, Elizabeth Bennett   84
Edwards, John   67
Education Officer   161–163, 167
Egan, Daniel   91
Elizabeth II, Queen   3, 151, 158, 176
Ellis, John   5
Ellis, Malcolm   18
Ellis, Thomas   21
Elyard, William   90
Endeavour   5, 6
Endlicher, Stephen L.   87
Etheridge, Robert   101
Eucalyptus fastigata   174, 176
Eucalyptus robusta   24
Eulomene   129
Evans, Surveyor George   21, 22, 30, 193
Evans, Obed David   158, 205
Exhibition, Intercolonial, Sydney (1870)   101
Exhibition, International, Sydney (1879)   105, 115, 203
Exhibition, London (1862)   95, 100
Exhibition, Paris (1855)   87, 95 (1867)   100, 202
Experiment Farm, Glen Innes   148
Eyre, Edward John   82
Farm Cove   ix, 8, 11, 13–16, 19, 21, 26, 44, 54, 80, 86, 97, 100, 122, 133, 138, 143, 151, 160, 164, 168, 169, 180
Fazzini, Antonio   205
Ferris, John Edward   xi, xiii, 167, 205
Fiaschi, Colonel Thomas   131
Ficus macrophylla   103, 105, 145, 152
Ficus virens (F. cunninghamii)   145
Fitz, Robert   15
Fitzgerald, Robert David   100
Fitzroy, Sir Charles Augustus, Governor   71–73, 79, 82, 85, 87, 90, 180
Flax   10
Fleischmann, Dr A.J.   149
Fletcher, Joseph James   116
Fletcher, Michael   166
Flinders, Lieut. Matthew   50, 165, 169
Flockton, Margaret   125, 126, 134, 139
Flood, Nicholas   21, 26, 193
Flora Australiensis   49, 99, 108, 110, 161

Flora of Australia   155, 160, 161, 164, 205
Flora of NSW   161, 163
Forbes, Justice Francis   43, 45, 61, 179
Ford, Neridah C.   155
The Forest Flora of NSW   125, 129, 141
Forsyth, William   118, 120, 130
Fowkes, Francis   11
Fowler, Verlie   204
Frame, David Ross   149
Francis, Robert   21
Franklin, Nathaniel   13
Fraser, Charles   26, 29, 30, 34–51, 84, 120, 169, 178, 182, 188, 195, 196
Friends Of The Gardens   xi, 161, 166, 167
Froggatt, Walter Wilson   116, 118, 126, 130, 147, 189
Furby, E.B.   148
Garden Island   19, 111, 148
Garden Palace   105–107, 115, 116, 171
Gardener's Magazine   48
Gardiner-Garden, Joy *see* Thompson, Joy
George III, King   3, 5–7, 27, 37
Gerrald, Joseph   13, 14
Gethsemane, Garden Of   2
Gill, James   21
Gipps, Sir George, Governor   62–64, 66–71, 193, 199, 200
Gladstone, William Ewart   71, 72
Glasshouses, *see* Conservatories
Gleeson, Gerald   xiii, 165
Glenelg, Charles, Baron   56, 63
Goddard, A.E.   129
Goderich, Frederick John, Viscount (First Earl of Ripon)   43, 50, 51, 53
Gold Discoveries   85, 87
Goulburn, Frederick   40
Government Farm   ix, 11–16, 20, 21, 54, 163
Govett, Surveyor William Romaine   178
Gowrie, Zara Eileen, Lady   147
Graham, Thomas   41, 50
Grahame, William C.   137
Grant, Alexander   120, 129
Grant, Charles R.G.   158
Grant, John   103, 104
Grant, William   135, 136, 140, 141
Great War, *see* World War I
Greenhouses; *see* Conservatories
Grevillea robusta   101, 172
Grewcoe, Ronald   160
Grey, Henry, Third Earl   72, 75, 77
Gritton, Charles   201
Guide to the City of Sydney   185
Guildford   26
Guilfoyle, Michael   90, 91, 93
Guano   70
Gymnasium, Juvenile   123, 124
Haffield, H.E.   104
Hall, Edward Smith   42, 45, 49
Hall, Peter   158
Hamilton, Arthur Andrew   131
Hamilton-Grey, Mrs A.M.   158
Hampden, Henry R.B., Viscount (Governor)   121
Handbook of the Flora of NSW   110

Harden, Gwen   164
Hardie, William   141
Hargraves, Edward Hammond   85
Harkins, Patrick   111
Harriet   30, 32
Harris, Surgeon John   15
Harris, Thistle Y. *see* Stead, Thistle Y.
Harvey, Professor William Henry   120
Harwood, George   89, 118, 122, 134
Havannah   84, 201
Hawkey, George Frederick   x, 143, 145, 146, 148, 149
Heath, Harry   156
Hedyscepe canterburyana   100
Heffernan, Martin   70
Henderson, Dr John   48, 188, 196
Henderson Lieut. John   180
Henessey, William   21
Henshall, William   37
Henslow, Professor John   72, 77, 200
Herbarium   ix, x, 5, 51, 53, 66, 82, 88, 89, 116–118, 120, 125, 126, 129, 130, 133, 139, 140, 144–150, 152–158, 160–163, 165, 166, 185, 187
Heroine   64
Heuckelum, Margaretta van   112
Heward, Robert   66
Hewlett, George   21
Hibiscus heterophyllus   21
Higgerson, John   26
Hoddle, Surveyor Robert   174
Hodgson, Christopher P.   179
Holman, William A.   137, 206
Hood, John   179, 180
Hooker, Sir William Jackson   48, 50, 52, 64, 66–73, 81, 87, 89, 92
Hopetoun, John Adrian Louis Hope, Seventh Earl (Gov.-Gen.)   123, 124
Hordern, Major Anthony   138, 140
Horrocks, Rev. Jeremiah   5
Horticultural Magazine   99
Hot-House, *see* Conservatories
Howea belmoreana   100
Howea forsterana   100
Howitt, William   181
Huntley, William   104
Hunter, Captain John, Governor   13, 15, 16
Hyde Park   18, 39, 57, 58, 78, 82, 87, 111, 134, 146
Hynes, Sarah   120, 130, 154
Illawarra   36, 37, 40
Illustrated Guide to Sydney   185
Illustrated Sydney News   104
Ingram, Keith   206
Investigator   50, 165, 169
Jacobs, Betty J.   163
Jacobs, Dr Surrey W.L.   163, 164, 167
Jameson, Surgeon R.G.   181
Jamison, Sir John   58, 63, 179, 198
Japanese Garden   xi, 160
Jenner, Miss A.M.   137
Jessop, William   183, 184
Jevons, William Stanley   177, 206
Job, Charles   52, 53
John Barry   38
Johnson, Dr Lawrence Alexander

Sidney ix–xii, 149, 154–156, 160–164, 167, 177, 187, 188
Johnstone, John 163
Jones, Valerie 155
Keele, T.E. 133
Keen, Gordon 188
Kelly, John 21
Kemp, George 96, 97
Kemp, Leonie M. 167
Kennedy, Edmund B.C. 82
Kennedy, William 21
Kerr, John R., Q.C. (later Sir) 154
Kew Bulletin 126, 204
Kew Gardens 5, 6, 16, 28, 29, 37, 50, 53, 62, 72, 77, 150, 158, 163, 167, 193
A Key to the Eucalypts 146
Kidd, James 53, 55, 61, 62, 67–72, 80, 91, 92, 100, 111, 200, 202
Kimber, William 15
King, Lieut. Philip Gidley (later Captain and Governor) 10, 15, 16, 168, 169, 192
King, Lieut. Phillip Parker (later Captain and Rear-admiral) 30, 32, 36, 37, 41, 58, 65, 69, 73, 91, 92
Kitchen, John 201
Kosciusko, Mount 120
Kunth, Carl S. 87
Lachlan Swamp 170, 171
Lady Castlereagh 34, 36
Lady McNaughton 57, 198
Lang, Rev. John Dunmore 181
La Perouse, Jean Francois Galaup, Comte de 8, 10, 11
Latrines 118, 121, 137, 144, 145
Lawson, Lieut. William 21
Lee, Alma T. 154, 155, 205
Leichhardt, Friedrich Wilhelm Ludwig 66–70, 74–76, 179, 199
Le Receveur, Abbé 11
Levy, Lewis Wolfe 150
Lewin, John William 22, 31
Lhotsky, Dr John 53
Library, Botanic Gardens 87, 92, 93, 103, 105, 117, 120, 121, 125, 143, 145, 146, 165, 166, 188–190
Library, Mitchell 126, 128, 129, 143, 189, 190
Lighting 96, 107, 118, 121, 122, 127, 132, 137, 151
Lindley, George 68
Lindley, Dr John 72, 73, 75–77, 87, 89, 97
Linnaeus, Carolus 5, 6
Liversidge, Professor Archibald 115
Loftus, Augustus, Lord (Governor) 106
Lophostemon confertus 105
Lord Howe Island 100, 108, 120, 163
Loudon, John C. 48, 87
Lucas, Arthur H.S. 120, 143, 203, 204
Lucas, Nathaniel 193
Macarthur, Edward 43
Macarthur, Elizabeth 47, 196, 197
Macarthur, James 91, 92
Macarthur, Sir William 58, 63, 67–69, 72–76, 81, 82, 95, 99, 179, 198
McBarron, E.J. 165
McElhone, John 108

McDonald, Alexander 13
McEwen, John 111, 118, 131
McGill, Walter 149
McGillivray, Donald J. 158
McGirr, John Joseph Gregory 136
Mackay, Dr James T. 77
Mackinnon, Ewen 131
Mclean, John 41, 43, 49, 52, 53, 55, 57, 198
Mcleay, Alexander 43, 47, 48, 50, 52–56, 58–60, 63, 67–70, 74, 179, 199
Macleay, George 58, 68, 69, 82, 90–92
Macleay, William 91
Macleay, William Sharp 66, 82, 90
Mclerie, John 96
Macquarie, Elizabeth 17, 22–26, 39, 181, 182, 184; *see also* Mrs Macquarie's chair and Mrs Macquarie's drive
Macquarie, Colonel (later Major-General) Lachlan (Governor) ix, 13, 16, Chapter 3, *passim*, 17–39, 169, 170, 178, 192–195
Maiden, Harrie 116, 129
Maiden, Joseph Henry ix, 26, 65, 84, 89, 107, 111–113; Chapter 6, *passim*, 114–141 (*birth*, 114; *education*, 114–116; *arrival in Sydney*, 115; *marriage*, 116; *family*, 116; *appointment as Director*, 117; *honours*, 137, 140; *travels*, 120, 122, 123, 129; *retirement*, 140, 141; *death*, 141, 143, 160, 165, 168, 174, 176, 186–189, 192, 193, 203
Maiden, Mary 116, 125, 126
Mair, Herbert Knowles Charles x, xii, 131, 154, 156–160, 176, 189
Mann, David Dickenson 15
Markett, Frederick 13
Marsden, Rev. Samuel 49, 52, 197
Martin, Sir James 149
Mary 52
Matra, James Maria 7
Mauritzon, Dr Johan 150
May, Valerie, see Jones, Valerie
Medway 75, 201
Meehan, Surveyor James 169, 170, 192
Melbourne, Royal Botanic Gardens ix, 183, 185, 191
Meredith, Louisa 179, 180
Mermaid 32, 37
Metrosideros 16
Meurant, Ferdinand 15
Mitchell, Sir Thomas 42, 56, 82, 88, 89, 198
Moctezuma, King 2
Monitor, see *Sydney Monitor*
Montefiore, Joseph 61
Moore, Alexander 105
Moore, Charles ix, 72, 73; Chapter 5 *passim*, 75–113 (*birth*, 77; *appointment as Director*, 72; *arrival in Sydney*, 75; *marriage*, 84; *travels*, 82, 84, 87, 95, 100, 104, 110; *honours*, 113; *retirement*, 112; *death*, 112, 113), 114, 116, 117, 120, 129, 134, 138, 151, 170–173, 180, 188, 200, 201, 203.
See also Committees of enquiry
Moore, David 77, 104
Moreton Bay Fig, see *Ficus macrophylla*

Morphett, Irene Eva 205
Mort, Thomas Sutcliffe 90, 104
Mould, John 18
Mrs Macquarie's chair 23, 25, 96, 118, 140, 177, 196
Mrs Macquarie's Drive/Road 23–25, 37, 184
Mueller, Baron Sir Ferdinand Jakob Heinrich von ix, 100, 108, 110, 111, 116, 117, 138, 149
Muir, see Moore
Mundy, Lieut.-Colonel Godfrey Charles 180
Murphy, Lionel, Q.C. (later Mr Justice) 154
Murray, Ruth 163
Museum, Botanic Gardens 105, 107, 120, 122, 125, 139, 145, 150
Museum, Technological 115–117, 144
Musgrave, Anthony 131
Myers, Francis 184
Myers, Sir Rupert 166
Nash, Richard W. 72
National Park 111
Nebamun 1
Nebuchadnezzar, King 1
Nepean, Under Secretary Evan 7, 10
Neville, John 101
Newman, Francis W. 67, 68
Nichol, John M. 141
Nichol, Richard 117, 141
Nicholas, Helen xii, 204
Nichols, George Robert 91
Nichols, Isaac 90
Nicholson, Dr Charles 66, 69, 82, 92
Noble, Dr Robert J. 148
Norfolk 57
Norfolk Island Pine, see *Araucaria heterophylla; see also* Wishing Tree
North, Alfred J. 131, 137
Nott, Roger B. 154
Nurseries, Private 26, 58, 60, 82, 85, 89–92
Nursery, State (Campbelltown) 105, 111, 112, 117, 118, 128, 131, 137, 143, 144, 170, 204
Nutgrass, see *Cyperus rotundus*
O'Brien, Cornelius 40
Opuntia 121, 122, 125
Orchid Collections 117, 129, 130, 140, 149
Orchids of NSW 149
Orphan Boys 47, 49, 53
Owen, Robert 56
Oxley, Surveyor-General John 22, 27–30, 33–35, 37, 39, 120, 194
Palace Gardens 107, 111, 117, 118, 120, 124, 128, 129, 138, 143
Palm Grove 12, 101, 110, 150, 156
Palmer, Commissary John 14, 18–20, 193
Papillio, Domenico 21
Parkes, Sir Henry 97, 117, 170–172
Parkinson, Sydney 5
Parr, William 29, 30
Parramatta 13, 28, 30, 36, 40, 41, 49, 52, 58, 116, 168, 169, 205
Paspalum dilatatum 138
Passey, Beverley xii, 205
Paterson, Lieut.-Col. William,

Lieut.-Governor 17, 43, 168, 169
Penfold, Arthur Ramon 144
Peron, Francois 16, 169, 192
Peters, Charles 112
Phillip, Captain Arthur, Governor ix, 7–10, 14, 15, 89, 121, 151, 177, 191, 192
Phillips, Sampson 21
Philosophical Society (Aust.) 41 (NSW) 100
Phormium tenax 10
Phragmites australis 35
Pickard, John 158, 162–164
Pioneers' Memorial 107, 147, 159
Piper, Captain John 169, 170
Pitman, Professor Michael George xi, xiii, 161, 166, 205
Plant-Houses, *see* Conservatories
Plants, Early Introduced 11, 13, 15, 16, 30, 41, 43–46, 48–50, 54, 56–58, 62, 93, 104, 111, 196, 197
Platypus 33, 99, 194
Playfair, George Israel 129, 131
Porter, George 59, 198
Powell, Dr Jocelyn M. 163
Price, Elizabeth 161
Prickly Pear, see *Opuntia*
Pringle, James M. 120
Public Reference Collection 166, 167, 187
Public Service Journal 126, 204
Purcel, Richard 201
Pyramus 67
Railway, Underground 109, 144, 148, 171
Railway Stations (Gardens) 111, 137, 174
Rainbow, William Joseph 131
Ralph Rashleigh 57
Ramsden, Abraham 15
Rattlesnake 61
Read, Daniel 37
Regulations (Botanic Gardens) 79, 80, 108, 109, 184
Rendall, Kenneth James 166
Rennie, James 68
Richardson, John 26, 36
Rider, Thomas 21
Robertson, Sir John 95
Robertson, Nasmith 67–71, 126
Rodd, Anthony 163, 166
Rodway, Dr F.A. 150
Rose Hill, *see* Parramatta
Ross, George David 144
Ross, Major Robert, Lieut.Gov. 10
Rossi, Francis Nicholas 55
Royal Botanic Gardens and Domain Trust vii, xi, xiii, 164–167, 176, 178, 187
Royal Epithet 3, 152
Royal Society 4, 5, 17, (NSW) 186
Royal Visits 123, 129, 130, 151, 152, 160, 165
Rumex hymenosepalus 108
Rupp, Rev. Herman Montague Rucker xii, 149
Ryan, James 201
Ryan, William 134
Saint John's Cemetery, Parramatta 13, 20, 49
Salisbury, Sir Edward 150, 156
Sand Erosion (Newcastle) 56, 87
Sassafras *see Doryphora sassafras*

Satellite Gardens 151, 156, 160, 161, 164, 165, 174–176, 187
Scientific Arrangement, see Arrangement Grounds
Scott, Helenus 68
Scott, Archdeacon Thomas Hobbes 43
Sea-Wall 82, 84, 85, 93, 94, 104, 184
See, John 125
Senecio linearifolius 174
Shakeley, Ned 26
Shanks, Dr Archibald 82, 83
Shea, John 15
Shepherd, Thomas 58, 202
Shepherd, Thomas William 90, 91, 202
Sheridan, James 21
Shipley 35
Simonetti, Achille 121
Sirius 7
Slim, Field Marshall Sir William, Governor-General 151
Sly, John 21
Smart, Thomas W. 90
Smilax australis 174
Smith, Henry George 116
Smith, James 13
Smith, John 71
Smyth, Surgeon Arthur Bowes 9, 191
Snodgrass, Colonel Kenneth 62
Solander, Daniel 5, 6, 126, 166
Southwell, Daniel 10, 192
Speedy 168
Spöring, Herman 5
Sprengel, Kurt 87
Stanley, Edward, Lord (later Earl of Derby) 66, 67, 69–71
Statues Committee 150

Stead, Thistle Yulette xii, 147
Steenbhom, Lionel Harold 150
Stevens, Hon. Bertram S.B. 144
Stewart, Nellie 151
Strickland, Sir Gerald, Governor 137, 206
Stuart, Alexander 108
Stuart, John, Third Earl of Bute 5, 6
Sturt, Captain Charles 58, 82
Sturt's Desert Pea see *Clianthus formosus*
Sugar Cane 88, 94, 103
Summer Houses 98, 99
Supply 7
Surry 41
Sydney, Lord 7–10
Sydney Gazette 42, 44, 45, 190
Sydney Mail 97, 202–204
Sydney Monitor 42, 44, 45, 190
Sydney (Morning) Herald 63, 65, 85, 126, 134, 147, 172, 179, 190, 199
Taylor, Sir George 156
Telfair, Charles 43
Telopea 163
Tetragonia tetragonioides 46
Theophrastus 2
Thetis 100
Thompson, Dr John Vaughan 59, 67, 198
Thompson, Joy 155
Thompson, Robert 201
Thomson, Edward Deas 59, 60, 62, 67, 77, 88, 94, 199
Thorne, Graeme 154
Thozet, Anthelme 94, 111
Tindale, Dr Mary D. xii, 150, 155, 156, 163
Tobin, Martin 67
Tomah, Mount x, 56, 95, 120, 156,

160, 161, 164, 165, 174–176, 187, 188, 206
Town and Country Journal 184, 190
Townsend, Joseph P. 181
Transportation, Cessation Of 68
The Trees of NSW 146, 149
Trollope, Anthony 2, 183, 184
Trust, see Royal Botanic Gardens and Domain Trust
Tucker, James 57
Tunks, William 97
Turner, Frederick 107
Turner, Rev. George Edward Weaver 81, 82, 84, 89–92
Tyrrell, Rt. Rev. William 75
The Useful Native Plants of Australia 114, 116
Valder, George 136, 143
Venus, Statue of 146
Venus, Transit of 4, 5
Vickery, Dr Joyce x, 149, 154, 156, 157, 163, 205
Victoria Lodge 143
Victoria Park 111
Visitor Centre 26, 115, 166, 188
Walker, Rev. James 84
Walker, William 21
Wallace, Dr Ben 166
Walls, Macquarie's 16, 18, 20, 24, 25, 101, 143, 147, 167
Ward, Edward Naunton 134, 140–145
Ward, Captain Edward W. 91
Ward, W.H. 144
Washington, John 132
Watermann, William 70
Watling, Thomas 10, 192
Watson, Maurice 151
Watson, Warwick xii, 151, 158, 159, 161–163, 176, 188

Wattle Day Movement 130, 146
Watts, Roy x
Watts, Rev. William Walter 129, 138
Wauch, Robert Andrew 59, 198
Webb, James 185
Wellesley, Arthur, Duke of Wellington 39, 43
Wentworth, D'Arcy 26
Wentworth, William Charles 21, 177
Wentworth Park 111, 131
Whitelegge, Thomas 137
Wickham, Captain John 65
Wilshire, James 91
Wilson, David 96
Wilson, Edwin xii, 165, 189, 192
Wilson, Rev. Francis R.M. 120
Wilson, Colonel Henry Croasdaile 57, 58
Windmills 19, 20, 193
Wishing Tree 14, 26, 27, 104, 147, 149
Woccanmagully ix, 8
Woodhart, Edward 69, 70
Woolley, Thomas 90
Woolls, Rev. Dr William 116, 120, 174
Worgan, Surgeon George 9, 11–13, 192
World War I 134, 135, 138, 146–148
World War II 146–149, 155, 156
Wran, Hon. Neville K. vii, xiii, 142, 154, 164–166, 176
Wynyard, Maj. Gen. Edward 81
Xanthorrhoea 98, 99, 155
Zeck, Emil 131
Zouch, Lieut. Henry 56
Zulaikha, Brian 166

Royal Botanic Gardens Sydney

DATE DUE

DUE DATE SUBJECT TO CHANGE
IF A RECALL IS REQUESTED

DEMCO, INC. 38-2931

OPERA HOUSE GATE

GOVERNMENT
HOUSE DELIVERY GATE

Government House

INNER
DOMAIN DEPOT GATE

YURONG GATE

Twin
Ponds

STREET